Praise for

The Biochar Solution

Reading like a detective story and marked by impressive
scholarship, Albert Bates' latest book has placed the biochar
solution and the vision of a truly regenerative agriculture
and settlement squarely in the center of the global crisis.
New historical evidence that climate is remarkably responsive
to human impacts had me gripping the edge of my seat.
The comprehensive and well-informed review of current
initiatives and technologies is a tour-de-force, and the grasp
of the global policy debate equally sobering. It is hard to
imagine a technical subject — compounded of organic
chemistry, archeology, rural economics, climate science, and
microbiology — presented with greater drama or clarity.

— Peter Bane, *Permaculture Activist*

In *The Biochar Solution*, Albert Bates demonstrates the flaws
of the story on which industrial civilization is based and offers
the living of a new story that will be created by changing our
relationship with the planet, and specifically its carbon element.
As a result of decades of experience, Bates is better equipped
than anyone I know to guide us in slowing climate change by
creating carbon-neutral cities and solidly sustainable agriculture.

— Carolyn Baker, Ph.D.,
author of *Sacred Demise: Walking The Spiritual Path of
Industrial Civilization's Collapse*

This book should be required reading for every policymaker, as well as everyone who eats food, breathes air, enjoys life and wishes to continue doing so. Bates has woven together a highly engaging interdisciplinary answer to climate change that draws on archaeology, history, ecology, chemistry, philosophy, and his vast and eclectic personal experience, a lively page-turner that blends clear-headed analysis with nuts-and-bolts advice. The Chinese symbol for crisis, he reminds us, is comprised of two words: danger and opportunity. He gives us both sides of that coin — enough danger to wake us up, but ample opportunity to emerge feeling hopeful.

— Tracy Barnett, multimedia travel journalist,
author and founder and editor of The Esperanza Project,
www.TheEsperanzaProject.org.

For things to remain the same, everything must change. Before I traveled to Copenhagen for the climate conference, a Benedictine monk asked me if I thought the survival of the human race was politically feasible. I have reflected on that question many times since then. As *The Biochar Solution* illustrates, climate change cannot be dealt with solely through scientific and economic means. Social and motivational transformation are essential components of the equation.

— Feargal Duff, Senior Advisor to the
Foundation for Economic Sustainability, Ireland

The
Biochar
Solution

Books for Wiser Living
recommended by *Mother Earth News*

Today, more than ever before, our society is seeking ways to live more conscientiously. To help bring you the very best inspiration and information about greener, more sustainable lifestyles, *Mother Earth News* is recommending select New Society Publishers' books to its readers. For more than 30 years, *Mother Earth* has been North America's "Original Guide to Living Wisely," creating books and magazines for people with a passion for self-reliance and a desire to live in harmony with nature. Across the countryside and in our cities, New Society Publishers and *Mother Earth* are leading the way to a wiser, more sustainable world.

The
Biochar
Solution

CARBON FARMING AND CLIMATE CHANGE

ALBERT BATES

Foreword by Dr. Vandana Shiva

NEW SOCIETY PUBLISHERS

Cataloging in Publication Data:

A catalog record for this publication is available from
the National Library of Canada.

Cover design by Diane McIntosh.
© iStock Roman Milert (Field)/ eyewave (charcoal)
Printed in Canada. First printing September 2010.

Paperback ISBN: 978-0-86571-677-3
eISBN: 978-1-55092-459-6

Inquiries regarding requests to reprint all or part of *The Biochar Solution* should
be addressed to New Society Publishers at the address below.

To order directly from the publishers, please call toll-free (North America)
1-800-567-6772, or order online at www.newsociety.com

Any other inquiries can be directed by mail to:
New Society Publishers
P.O. Box 189, Gabriola Island, BC V0R 1X0, Canada
(250) 247-9737

New Society Publishers' mission is to publish books that contribute in funda-
mental ways to building an ecologically sustainable and just society, and to do so
with the least possible impact on the environment, in a manner that models this
vision. We are committed to doing this not just through education, but through
action. Our printed, bound books are printed on Forest Stewardship Council-
certified acid-free paper that is **100% post-consumer recycled** (100% old growth
forest-free), processed chlorine free, and printed with vegetable-based, low-VOC
inks, with covers produced using FSC-certified stock. New Society also works to
reduce its carbon footprint, and purchases carbon offsets based on an annual audit
to ensure a carbon neutral footprint. For further information, or to browse our
full list of books and purchase securely, visit our website at: www.newsociety.com

Library and Archives Canada Cataloguing in Publication

Bates, Albert K., 1947-
    The biochar solution : carbon farming and climate change / Albert Bates.

Includes index.
ISBN 978-0-86571-677-3

    1. Charcoal. 2. Soil amendments. 3. Carbon sequestration.
4. Agriculture--Environmental aspects. I. Title.

TP331.B37 2010   662'.74   C2010-905765-1

MIX
Paper from
responsible sources
FSC™ C016245

NEW SOCIETY PUBLISHERS

Contents

Acknowledgments

STEVE PAYNE, ERIK ASSADOURIAN AND BRUNO GLASER got me started writing on this subject; Johannes Lehmann, Stephen Joseph, David Manning, John Gault, Jim Amonette, Julie Major, and many others were very patient in answering my questions about biochar; Toby Hemenway and Peter Bane inspired me with their grasp of soil chemistry at a talk they gave at the Lama Foundation; Mary Ann Simonds and Daniel Kittredge introduced me to carbon farming; Darren Doherty, Eric Toensmeier, Brad Lancaster, Joel Salatin, and several others came to Tennessee to teach the first practical carbon farming course, and Greg Landua, Patrick Gibbs, and Ethan Roland showed me how it was done here in a compacted horse pasture; Chris Nesbitt invited me to teach permaculture at his agroforestry research farm in Belize; Jeff Wallin, Hugh McLaughlin, Jim Fournier, John Miedema, Danny Day, and Lopa Brunjes helped me through the biochar industrial processes; Ron Larson, Paul Anderson, Erich Knight, Folke Günther, David Yarrow, Ianto Evans, David Friese-Greene, and Kelpie Wilson helped with stove designs; Frank Michael and Nathaniel Mulcahy gave me biochar stoves to experiment with and checked my arithmetic; Leo Principe and Vanessa Marino housed and transported me by land, sea, and air all over Amazonià; Newton Falcão, Christoph Steiner, and Charles Clement introduced me to many other Brazilian terra preta scientists and accompanied us to dig sites in Açutuba; Peter Harper introduced me to Pliny Fisk and David Orr; all of whom were very helpful; Elaine

Ingham, Geoff Lawton, Lilian Rebellato, Adam Posthuma, and others I have already mentioned double-checked many of my facts; Davie Philip, Bruce Darrell, Ben Whelan, and Graham Strouts gave me the insider's tour of The Village; David Fleming, Richard Douthwaite, and Corinna Byrne helped me fathom the economics of carbon trading; Ross Jackson, Feargal Duff, and Maurice Strong spent many breakfasts and dinners in Copenhagen patiently dissecting the goings-on; David Haenke walked me through the Alford Forest and told me the back-story I had never known about my old friend Leo Drey; Susan Flader updated the Pioneer Forest history; Sandor Katz introduced me to bacteria, Ronald Nigh to milpas, Scott Horton to chinampas, and Jan Garrett and K.C. Das to organic no-till; what I know of biodynamic agriculture I learned from Jeff Poppen, Bob Kornegay, Jan Bang, and Declan Kennedy, and from visits to many Camphill and biodynamic farms in the United States, Germany, England, Iceland, Portugal, and Scotland; those who helped me understand climate science could fill several pages so I won't even begin, but I must acknowledge the time taken by Stephen Schneider, Ross Gelbspan, Susan Solomon, and Jim Hansen; Chris and Judith Plant, Heather Nicholas, Sue Custance, Ingrid Witvoet, Stephanie Mills, and Gayla Groom were my indefatigable sounding boards, editors, and production managers; and without all of the above, this book would have been far less interesting.

Foreword

by Dr. Vandana Shiva

Cultivating the Future

Fertile soils rich in organic matter are are our best insurance against food insecurity and climate vulnerability. Soil is a major store of carbon, containing three times as much carbon as the atmosphere and five times as much as forests. About 60% of this is in the form of organic matter in the soil. The principal component of soil carbon is humus, a stable form of organic carbon with an average lifetime of hundreds to thousands of years.

Soil organic matter determines much of the soil's quality, as it "is an important substrate of cationic exchange, is the warehouse of most of the nitrogen, phosphorous and sulphur potentially available to plants, is the main energy source of micro-organisms and is a key determinant of soil structure", (J. Ewel, "Designing Agricultural Ecosystems for the Humid Tropics," *Annual Review of Ecology and Systematics*, 17:245-271)

A pioneer of organic agriculture, Sir Albert Howard defined fertile soils as follows:

a soil teeming with healthy life in the shape of abundant microflora and microfauna, will bear healthy plants, and these, when consumed by animals and man, will confer health on animals and man. But an infertile soil, that is, one lacking sufficient microbial, fungous, and other life, will pass on some form of deficiency to the plant, and such plant, in turn, will pass on some form of deficiency to animals and man.

The millions of organisms found in soil are the source of its fertility. The greatest biomass in soil consists of microorganisms, fungi in particular. Soil microorganisms maintain soil structure, contribute to the biodegradation of dead plants and animals, and fix nitrogen. Their destruction by chemicals threatens our survival and our food security.

Industrial agriculture treats soil as an empty container for industrial fertilizers. After World War I, manufacturers of explosives, whose factories were equipped for the fixation of nitrogen, had to find other markets for their products. Synthetic fertilizers provided a convenient conversion for peaceful uses of war products. Howard identified this conversion as closely linked to the "NPK mentality" of chemical farming:

> The feature of manuring of manuring of the west is the use of artificial manures. The factories engaged during the Great War in the fixation of atmospheric nitrogen for the manufacture of explosives had to find other markets, the use of nitrogenous fertilizers in agriculture increased, until today the majority of farmers and market gardeners base their manurial program on the cheapest forms of nitrogen (N), phosphorous (P), and potassium (K) on the market. What may be conveniently described as the NPK mentality dominates farming alike, in the experimental stations and in the countryside. Vested interests entrenched in time of national emergency, have gained a stranglehold.

After the Wars, there was cheap and abundant fertilizer in the west, and American companies were anxious to ensure higher fertilizer consumption overseas to recoup their investment. The fertilizer push was an important factor in the spread of new seeds, because wherever the new seeds went, they opened up new markets for chemical fertilizers.

This is how industrial agriculture was introduced as the "Green Revolution" in India in the 60's. However, replacing soil fertility with chemical fertilizers was neither green nor revolutionary. It was a recipe for destroying soils, eroding food security and increasing green house gases which contribute to climate change. (Vandana Shiva, *Violence of the Green Revolution*, Zed Books, 1989)

However, the myth of the green revolution continues. On September 6, 2010 *Time* magazine stated in its cover story on the real cost of organic food, "Norman Borlaug, the so called father of the green revolution, who nearly doubled wheat yields in Pakistan and India in the 60's via a combination of high yield plants and fertilizer use, is often credited with saving one billion lives."

This account is false on many counts. Firstly, the so called "high yielding varieties" are in fact "high response varieties," engineered to withstand high doses of chemicals. Secondly, the increase in wheat production, which is assigned to chemicals and chemically adapted seeds, can be accounted for by any increase in land under wheat cultivation, and any increase in water provided for irrigation. Thirdly, high-cost external input agriculture is the *reason* for hunger. It has not saved a billion lives.

What the green revolution narrative ignores is decline in overall output, increase in costs of cultivation, and the destruction of the soil. Food security rests on soil building, not on poisoning the soil with toxics and burdening farmers with debt. Over the past decade, 200,000 farmers have committed suicide in India due to indebtedness resulting from high cost seeds and chemicals.

Building soil means building the soil food web in all its diversity and complexity. We need to build living soils because they are the very source of life. We need to build living soils because they provide diverse and multiple ecological services, including conservation of water and maintenance of the hydrological cycle. We need to build living soils because they are the basis of food security. And we need to build living soils because they provide climate resilience.

Living soils grow from living carbon. And living carbon is the result of the process of photosynthesis. As Sir Albert reminds us "life is maintained by the sun's energy and the instrument for intercepting this energy and turning it to account is the green leaf.....".

The green leaf, with its chlorophyll battery, is therefore a perfectly adapted agency for continuing life. It is, speaking plainly, the only agency that can do this, and is unique. Its efficiency is of supreme importance. Because animals, including man, feed eventually on green vegetation, either directly or through the bodies of other animals, it is our sole final source of nutriment. There is no alternative supply. Without sunlight and the capacity of the earth's green carpet to intercept its energy for us, our industries, our trade, and our possessions would soon be useless. It follows therefore that everything on this plant must depend on the way mankind makes use of this green carpet, in other words on its efficiency. (Sir Albert Howard, *Soil and Health*, University Press of Kentucky, pp20-21).

So for building soil we need to increase the green cover on the planet, both in forests and on farms. On our farms we need to shift from chemical and fossil fuel intensive monocultures to biodiversity and biodiversity intensive systems that multiply the production of living carbon with all the nutrients needed by the soil, plants and animals (including humans).

In *The Biochar Solution* Albert Bates walks us through the history of sustainable farming practices, the climate crisis, and the role of organic farming and building organic matter in soils in mitigation and adaptation to climate change. I fully endorse his vision of the world evolving into a garden. But I would like to sound a word of caution.

By shifting our concern from growing the green mantle of the earth to making charcoal, biochar solutions risk repeating the mistakes of industrial agriculture. The reductionist NPK mentality is replaced by a reductionist carbon mentality. The false assumption that soil fertility comes from factories is maintained. Earlier it focused on factories producing NPK, now it focuses on industrial production of biochar.

Just as industrial agriculture and the green revolution forgot about life, the biochar solutions are ignoring life with their carbon preoccupation, an example of what I have called the "Monocultures of the Mind".

We need to remember that calcium and magnesium, iron and copper, the Mychorrizae and the earthworm are also part of the soil's life, not just carbon. Above all we need to remember that carbon is fixed by the chlorophyll molecule in the green leaf of plants, not during the pyrolysis used to produce biochar.

The future cannot be built on the basis of knowledge that comes from a reductionist, fragmented, mechanistic world view. It cannot be built on the external input model of industrial agriculture.

To cultivate the future, we need to cultivate life in the soil. We need to cultivate the humility that the soil makes us, we do not make the soil, and we can only serve her processes of making life.

— Dr. Vandana Shiva
September 2010

Introduction

We are stardust, we are golden,
We are billion-year-old carbon,
And we've got to get ourselves back to the garden.

— Joni Mitchell

I N *OUR FINAL HOUR,* Sir Martin Rees tells the story of wandering into an antiquarian bookstore and browsing the old science fiction magazines from the first decades of the 20th century. None predicted nuclear energy or nuclear weapons, antibiotics or organ transplants, cheap air travel, GPS satellites, transistor radios, or iPhones. Rees observed that we are very likely as unaware today of what the world will be like in 50 years as science fiction writers of a century ago were unaware of what our world is like.

Some things we know.

We have known for more than a century now, since the predictions of Svante Arrhenius in 1896, that a doubling of atmospheric carbon could bring a 5° C rise in global temperature. But in the 1970s, most people who studied these things said the temperature was rising very slowly, only a half-degree per century, and we had time.

We got that wrong.

The temperature change in a single day is usually several degrees, even at the equator, so the thought of a four-degree change is not particularly frightening to most people. But to change the global average by even one degree requires heating an enormous volume of ocean water, an enormous volume of atmosphere, and an enormous volume of land surface. We are only about three degrees warmer today, on a global average, than we were 20,000 years ago, when there was a mile-thick mantle of ice over Manhattan. Four degrees more, which is where we may find ourselves in a few decades, is a big deal.

Scientists who came to this realization in the 1970s and 1980s became sufficiently alarmed to prod national governments to convene study groups and to institute the Intergovernmental Panel on Climate Change at the United Nations.

In 2007, the IPCC released its fourth summary report, and its warnings were unequivocal and dire. Earth is overheating. Species are being lost at a rate several thousand times faster than the historic average. Humans as a species are now in peril of extinction, by our own hand. Moreover, while the IPCC did not say it in so many words, the possibility cannot be excluded that we will take all other species of life with us when we go, and leave a desert world, parched clay dust blowing across the landscape under oven-like temperatures, oceans too toxic to support even microbial life forms. Worse, it is all happening at breathtaking speed, much too fast for our political systems to react, and could be irreversible before the 21st century is half out of its hourglass.

When you consider the extreme improbability that life could arise on this tiny blue outpost in space, the irony of our suicidal trajectory is even more stark. Earth resides in the Goldilocks zone (not too hot, not too cold, just right) of 0.95 to 1.15 astronomical units from a central, medium, yellow star; planets orbiting large, hot, or binary stars do not exist long enough to allow complex life to evolve. Earth has an orbit that steers wide of the high radiation levels at the galactic center and in its spiral arms. Earth has an iron core that generates a protective magnetosphere, ionosphere, and ozone shield. Were it not for Jupiter's size and gravitational pull, Earth would be bombarded by space debris. Were it not for our Moon's gravitational field, loose asteroids getting past Jupiter and the other planets would impact us with unsettling regularity.

All these protections played essential roles in deciding whether a magical assembly of mineral salts in the Archean sea could throw an envelope around itself and have children, or whether a live spore drifting outward in time and space from the Big Bang could take root on our world, survive, and thrive. We had to pass tests like these just to have water in a liquid state. It took water in liquid form to permit evolution of our multicellular organisms. A few major extinction events occurred, and then we had the Cambrian explosion, the Chicxulub impact, the die-off of the dinosaurs, and the ascent of the mammals. Only over very long periods of geological and climate stability could we have evolved our elaborate metabolism, our long in-utero gestation, a childhood lasting more than 25 percent of the average life span, and a holographic brain. Combine those off-chance developments with unusual eye-hand coordination, a nuanced vocal apparatus, and the 10-digit capability of formulating abstractions with mathematical precision, and you get video games — the teasing of our sudden-action retinal reflex — begetting artificial intelligence circuits in virtual realities where a protagonist may spontaneously ask, "Where did I come from?" and "Why am I here?"

There are many people writing and talking about extinction now, and about *solastalgia,* the psychic or existential distress caused by environmental change. Artists, musicians, dancers, and screenwriters are wringing profound solastalgia from the recesses of our collective subconscious and spinning its trajectory out into the light where we can get a better sense of it. Solastalgia does come with an antidote: ecological restoration. We can become a *garden planet.*

My own experiments with biochar began with a crude drum kiln. I made charcoal at low temperatures (300–600 °C) and then left it to marinate in urine from a pissoir outside my bedroom, combined it with compost, and shoveled it into raised garden beds. The results were immediately successful, and I have continued the practice for some years now, extending it to fill shallow trenches around trees in my landscape and any holes I dig for new trees. I am now making "designer chars" with specially brewed compost teas.

In 2008, I persuaded the editor of the Australian magazine *Organic Gardener* to send me to Newcastle, England, to the Second Conference of the International Biochar Initiative. It was appropriate that the conference met in Newcastle, birthplace of the fossil fuel era. Where better to talk about taking carbon from the atmosphere, turning it into coal, and burying it?

I went to Newcastle with a healthy skepticism — was biochar some sort of greenwash for accelerating deforestation, a palliative for advocates desperate for solutions to the climate crisis — or was there more to it? I came away a convert, my mind racing in a million directions; I had to hold myself back from evangelizing this new Gospel According to Biochar.

Biochar is a way of taking carbon directly from the atmosphere and thereby reversing the past two centuries of anthropogenic emissions, taking us from 390 parts per million by volume and all that portends — melting icecaps, spreading deserts, super-typhoons — to 350, 300, 250, and lower.

Biochar is charcoal, a cellulosic material that has been pyrolyzed — fired in a low-oxygen environment such as a kiln so that everything but the carbon has been burned off. As pure charcoal, it is hard (in Japan they make xylophone keys from it), black, and largely devoid of any nutrient value. It can produce relatively smokeless heat by being burned further in an oxygen-rich environment, which is why it is valuable for cooking and heating in many parts of the world.

A more significant value, to our ailing planet, is biochar's unique quality as a soil conditioner. Biochar is like a coral reef in the soil. If it is turned in a nutrient pile (any compost will do) and then tilled into the ground, it immediately becomes colonized by soil microbes, much in the way coral reefs are populated by all manner of marine life. The

Biochar Terminology

To avoid confusion, I have adopted a nomenclature convention developed by Johannes Lehmann and Stephen Joseph for a book they edited, *Biochar for Environmental Management* (2009). When I say "biochar," I mean charred (pyrolyzed) organic matter intended to be applied to soil in farming or gardening or for biological environmental remediation. When I say "charcoal" or "char," I mean charred (pyrolyzed or torrified) organic matter that might become biochar but might also be used as fuel, a filter, a catalyst in iron-making, or in any other industry or art. Biochar does not include charred plastics, tires, unsegregated landfill wastes, or other non-biological material; those are items that should be handled gingerly, if at all. Biochar also does not include "black carbon," which can be soot, graphite, or a broad range of oxidation products.

microbes attract fungi, which connect to the roots of plants, carrying nutrients from the reef to where they will do the most good.

Besides stimulating the health of the soil, the biochar provides a reservoir and conduit for soil moisture, soaking up water from over-saturated areas and giving it back to dry areas. One gram of charcoal (a piece about the size of a pencil eraser) has a surface area of 1000 to 2500 square meters (about the size of a house) because of all the micropores.

One take-home point from the Newcastle conference was that there is so much money to be made from burying this material that a veritable gold rush is ramping up, with entrepreneurs jumping in to patent processes and brand trademarks, con artists selling get-rich-quick schemes to gullible seniors, and venture capital firms salivating over the fortunes in carbon futures to be traded on global exchanges.

It is important to acknowledge that biochar is not a cure for addiction to industrial growth at any cost. Nor will changing agricultural practices be enough, alone, to reverse our planet's death spiral or to save endangered species, including ourselves. What needs to happen is a new approach to living within, rather than against, nature. We have to recover the indigenous wisdom that describes that path, what the Haudenosaunee call "the original instructions." We need to control our population size, our appetites, and our wasteful ways.

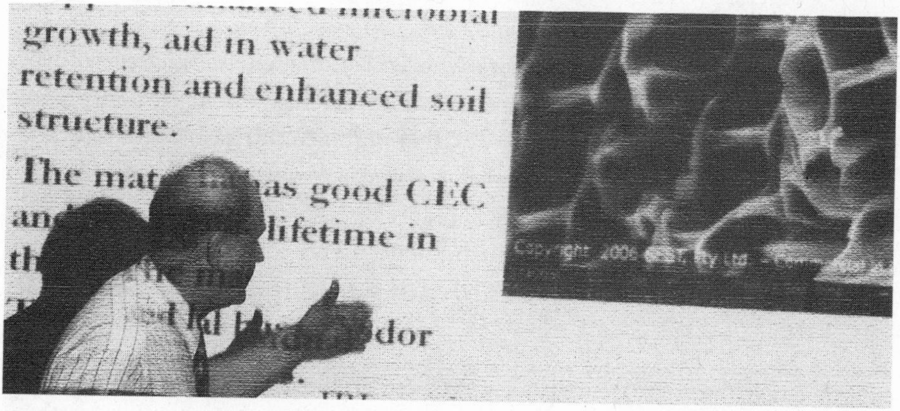

Fig. 1: *When cellulose is pyrolyzed, it forms pores, many of which may have pitted walls, or pores within pores.*

While researching this book, I had a chance to ask Elaine Ingham, a soil scientist who is best known for her discoveries in the microbial world of the soil food web, what she thought of biochar. It scared her, she confided.

I thought at first that she had been influenced by Biofuelwatch, one of biochar's most vocal critics, but as I probed her response, it turned out that she had good reason to be frightened. Biochar is too powerful, she told me. Once the industrial complex, with its credit markets, government incentives, and subsidies to farmers gets up and running, biochar could become a juggernaut, pushing the soil-atmosphere carbon balance into an overcorrection and ushering in a rapid-onset ice age.

"We don't need it," she said. "Just good soil-building practices could cancel out global CO_2 emissions and balance the atmosphere of the whole planet."

"Without biochar?" I asked.

"Without biochar."

This view is not uncommon. In fact, it was held by many of the soil scientists and organic farmers I met while researching this book. However, I have begun to question the idea that biochar is unnecessary. It is possible that both carbon farming and biochar combined may not be enough to bring global climate back from the brink of catastrophic change. To find our way out of the present dilemma, even more far-reaching change of practices will be needed, at a pace and scale that is unprecedented in our history.

I know there are readers who will immediately disagree with me about this. Many I respect insist that biochar is a hoax, that organic farming cannot feed the world, that permaculture or biodynamic practices are superstitious, or that agroforestry will never justify the land required for it. All I can ask is that people suspend judgment until they have read the book. After that, I may still not have won any arguments, but perhaps we will have begun a discussion.

BOOK I

LOSING THE RECIPE

*Any sufficiently advanced technology
is indistinguishable from magic.*

— Arthur C. Clarke

Although a population of seven billion people is by no means sustainable, given Earth's finite resources, we are now at a remarkable point in time. If talent is a genetic lottery, we are, at this moment, blessed to share our life spans with any number of Leonardos, Einsteins, Galileos, Ramanujans, and al-Khwārizmīs, alive and thinking. All of that collective brainpower — were it to survive our schools, civil conflicts, social injustices, and the other vicious obstacles we place in its path — might even equal the challenge now laid before us.

Moreover, there may already be a better path to guide our brightest minds, one that has been scouted and found useful, quick and safe. Our choice as a global civilization is to stay with the path we are on — one that turns forest and farm to salty deserts — or to try a different path — one that was widely practiced in nearly half the world, and then tragically lost. If our fates can realign, we might get back to where we once belonged.

The Roots of a Predicament

O N THE 10th OF NOVEMBER, 1619, while camped with the Habsburg army at Neuburg on the Danube, René Descartes had, in one night, three dreams, which he interpreted at the time, even before waking, to be revelations from the Spirit of Truth, reprimanding him for the sins of his youth but extending guidance for his future life.

In his third dream, the Spirit came to Descartes and said, "Conquest of nature is to be achieved through number and measure." That was the beginning of the Cartesian way of dissecting the natural world — one of mechanics, formulae, and, eventually, human design.

We might ask, however, if the words of the Spirit were intended not as a gift to Descartes, but as a grim prediction.

Approximately 400,000 years ago we, the two-leggeds, started using fire. A seasonal hunting camp called Terra Amata, in the South of France, dates from that period and shows what our lives were like.

The Terra Amatans would get up in the morning and go to work. For some, work was getting together with friends or relatives and heading out to hunt and fish. For others, it was gathering kindling and driftwood from the beach or looking about for edible roots, berries, eggs, mushrooms, and leaves in the nearby forest. The older children tidied up their infant siblings, swept the dirt floor in the sapling and skin shelter, put on some tea, or set about scraping leather, sewing, playing pebble "marble" games, or taking target practice with fist-sized rocks.

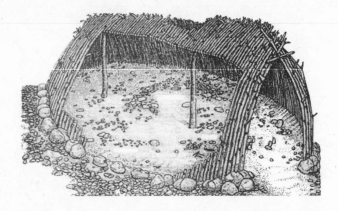

Fig. 2:
Terra Amata,
shown here in
artist's recon-
struction,[1] is
the earliest
known hominid
settlement.

As the Sun sank lower in the sky, the hunters and fishers would return with supper. There would be a large common meal, and people would retire to the shelter to sleep. If it were cold or raining, they would have the fire inside, and the hunters might stay home the next day to knap flint blades and fashion bone hooks and vine-rope nets, or to start work on some warm fur boots for the next winter. Kitchen wastes and humanure were disposed of away from the shelter. The cooking, knapping, and bone-whittling never took place in the sleeping area, lest someone awaken with a sharp shard of flint stuck in his or her backside.

The use of fire, sharp points, spears, and throwing sticks made it possible for Terra Amatans to keep animal predators at bay, vary what they ate, and set up their camps in different climates. As the seasons changed and the game patterns with them, the group would pull up stakes and move to another camp. The next year, when the time came again, they would return to the camp at Terra Amata and rebuild the shelter.

Nearly all of human history resembled this lifestyle. That's a remarkable thing, so I'll say it again.

Nearly all of human history resembled this lifestyle.

For the past 400,000 years, save about the last 11,000 or so, we two-leggeds did not plant crops or modify ecosystem dynamics in any enduring way. Whatever Gaia offered us we received gladly. We were always careful to give back in kind and we were part of, not apart from, the ecosystem.

Today we tend to look upon this way of living as very hard, even precarious, but most of the evidence we have from archeology or from "primitive" cultures studied by anthropologists suggests the contrary is true. In these cultures, life was usually stable and secure, and in some cases health and diets were better than in later cultures. The hunter-gatherers who were displaced by pastoralists and soil-tillers were not eliminated because they were unhappy or unsuccessful. In most cases, they just lacked comparable weapons, aggression, or tactical acumen.

Within a few thousand years of *Homo sapiens'* migrating out of Africa, many large game populations, including most of those hunted by *Homo neanderthalensis*, went extinct. Neanderthals, being bigger and heavier than our species, needed to consume 5000 kilocalories (kcal) of food energy daily, and did not eat plants. The smaller, smarter *Homo sapiens* needed only 3000 kcal, were much better hunters, gathered and ate plants, and had a broader arsenal of tools. The Neanderthals' Energy Return on Investment (ERoI), favorable for 400,000 years, suddenly turned unfavorable. When the two species met on the plains of Europe, it was no contest. *Homo neanderthalensis* disappeared, along with nine genera of kangaroo, a giant tortoise the size of a Volkswagen, and 33 types of large mammals.

About 100,000 years ago, our ancestors began befriending wild dogs; it was less a master-servant relationship, more a partnership of like interests, but gradually humans became involved in breeding dogs and training puppies. The dogs were smart and had better hearing, smell, and sight than we did, so they became patrolling scouts when the group was on the move. We made similar alliances with horses, sheep, goats, pigs, and rabbits.

Manipulating game animals to make them easier to catch was a no-brainer for a sharp team of hunters with plenty of time on their hands. We used fire to drive animals to places where they could be more easily harvested. We discovered that often in the post-fire vegetation there was a sudden profusion of berries, plants with bulbs, and mushrooms. We may have also observed that fire could make trees produce more nuts, or nuts produce trees where that kind had not emerged before. With this pattern in mind, we began employing "patch disturbance," or swidden agriculture.

Still, the ethic of keeping the balance of nature was very strong, as were the penalties for forgetting it. While our ancestors may have burned land to rejuvenate it, and to hunt, they were still highly adaptive, rather than manipulative.

In the Fertile Crescent 9500 years ago, we tamed the wild predecessors of many of today's farm animals; we domesticated grains and made bread and beer. With a year-round supply of protein, carbohydrates, and fats, we settled down and began to grow larger populations. The same process repeated itself in the Yellow River Valley of China, turning rice, millet, soya, and chickens into ever more human protein. The pattern repeated again in the Americas with squash, beans, corn, potatoes, tomatoes, llamas, saltwater fish, and cuy, the Andean guinea pig.

The herders and farmers that displaced the hunters and gatherers (or the "takers" versus the "leavers," to use Daniel Quinn's terms), set themselves a snare that any hunting society could have seen, but only recently have we, the agriculturalists, begun to realize it is around *our* legs, firmly knotted.

As we grew complex societies, built great cities, and reshaped ecosystems to produce greater yields of our favorite crops, we interlinked our commerce to make localities less subject to the vagaries of nature. However, once all commerce was linked — and was sustained less by renewable inputs and more by the nonrenewable, easily depleted kinds — we had unknowingly set the trap. It was sprung by a combination of energy peak and decline, depletion of a wide spectrum of nonrenewable resources, our *global* population growing beyond carrying capacity, and nowhere on Earth left to turn.

What is worse, it is now nearly impossible for us to revert to the older, long-since-supplanted, resilient food-producing systems that sustained us through 98 percent of human history. Even if we could wave a magic wand, reduce our population by 90 percent, revolutionize agriculture so that it no longer requires 10 calories of fossil fuels to produce each food calorie, reverse the skewed ratio between food providers and the other members of our cultures, and go back to living within nature's bounty — Mother Nature herself has moved on. What looms in the depths of the financial, food, and climate crises now unfolding is an almost unimaginable degree of hardship and starvation, if not annihilation of our species — and possibly all others, too.

Sombroek's Vision

W IM SOMBROEK WAS BORN IN HEILOO, the Netherlands, in late August 1934. It was not the best time to be born, because it meant that when he was six his country was invaded by the *Wehrmacht* (war machine) of Adolph Hitler, and little Wim's scenic 12th-century village would be an occupied theater of battle for the next five years.

And so, at the age of 10, he lived through the Dutch famine of 1944–1945 — the *Hongerwinter*. It was then that his father taught him the gardening skills that kept Wim and his family alive. His father toiled daily on a small plot of land made rich by generations of Sombroeks, and darkened by ash and cinders from their small home's wood heater.

Later, while working on his master's degree in soil science from Wageningen University, Wim traveled with an FAO/UNESCO (Food and Agriculture Organization/United Nations Educational, Scientific, and Cultural Organization) group to Brazil, where he observed the mysterious dark earths, or "terra preta," of the Amazon. It struck him that the rich earths resembled the *plaggen* soil from Holland that he knew as a child. When he looked at samples under the microscope, he confirmed the presence of partly charred cinders. Carbon dating showed the cinders were 4000 years old. Pottery shards found in the same layer with the cinders suggested the charcoal was of human origin.

In 1963, Sombroek was awarded a Ph.D. on his thesis of "Amazon Soils." He published his theories of the human origins of terra preta in book form in 1966; *Amazon Soils* sparked great interest in the soils and ecology of Amazonia, and further research began immediately in institutions of higher learning all over the world.[1]

In 1989, Sombroek traveled to Brazil again; in 1992, he published his first work on the potential of terra preta as a tool for carbon sequestration. When he submitted that piece for publication, he probably already understood what he had done, although few others at that time could have shared his insight. He had just possibly saved all life on Earth.[2]

Conquistadors

A YOUNG MAN BORN IN THE CITY OF TRUJILLO, SPAIN, around the year 1504 decided, or perhaps it was decided for him, that the life of the cloth was his calling. After schooling in the Dominican cloister in Trujillo, he took the name Gaspar de Carvajal. Towards the end of 1536, having proven himself a wise and scholarly friar, he was dispatched by his order, together with eight of his companions, to sail to Peru. Their mission, commanded by King and Emperor Carlos V, was to "with the evangelical word and the Cross of Christ, seed within the native race the message of consolation and resignation."

Fortune favored Father Gaspar, for his voyage to the Americas was dangerous, and once there, his assignment could have been very cruel or thankless. That was not to be his lot, although suffering and fear would seldom be very far away.

Crossing the Atlantic in a wooden barque, the group of friars passed around the coasts of Brazil and Argentina, through the harrowing Straits of Magellan, and up the Pacific coast of Chile to Peru, where they were received by the Provincial Vicar in Lima and there remained in service to the Conquistador Francisco Pizarro until 1538.

Father Gaspar distinguished himself through his service and leadership and was named the leader of his mission. Three years later, the Vicar was asked by subordinants of Gonzalo Pizarro, Francisco's younger half-brother in Quito, to recommend a suitable friar for an expedition inland in search of cinnamon; he recommended Carvajal.

A few years earlier, a 16-year-old conquistador named Francisco de Orellana had joined the fleet of ships commanded by his uncle, Francisco Pizarro, as it departed Spain. Although Ferdinand Magellan had by then charted the secret straits at the tip of Patagonia (improving on charts of Chinese origin obtained, it has been suggested, by Carlos V from the Vatican[1]), Pizarro's expedition did not sail to the Pacific. Instead, they landed on the Atlantic coast of what is now Nicaragua and passed by land to Mexico, where Hernán Cortez provisioned them with new ships. Pizarro then explored by sea southward to the Empire of the Inca.

By all accounts, Francisco Pizarro and his three brothers were among the most corrupt, brutal, and ruthless conquerors of the New World, committing unrestrained atrocities to "seed within the native race the message of consolation and resignation."

In 1541, as Hernando de Soto reached the Mississippi River, Sultan Suleiman the Magnificent sealed off the Golden Gate in Jerusalem, and Gerardus Mercator constructed his first globe, Gonzalo Pizarro was appointed governor of the northernmost of the four parts of the Incan empire. The rumor of El Dorado reached him from the 1537 expedition of conquistador Gonzalo Jiménez de Quesada into the Boyacá highlands of Colombia. "El Dorado" was actually used by Jiménez to describe a person — the "golden king" of the Muisca nation — rather than a place. In 1638, Juan Rodriguez Troxell wrote this account in a letter addressed to the cacique, or governor, of Guatavita, near present-day Bogotá:

> The ceremony took place on the appointment of a new ruler. Before taking office, he spent some time secluded in a cave, without women, forbidden to eat salt, or to go out during daylight. The first journey he had to make was to go to the great lagoon of Guatavita, to make offerings and sacrifices to the demon that they worshipped as their god and lord. During the ceremony that took place at the lagoon, they made a raft of rushes, embellishing and decorating it with the most attractive things they had. They put on it four lighted braziers in which they burned

much *moque,* which is the incense of these natives, and also resin and many other perfumes. The lagoon was large and deep, so that a ship with high sides could sail on it, all loaded with an infinity of men and women dressed in fine plumes, golden plaques and crowns. . . . As soon as those on the raft began to burn incense, they also lit braziers on the shore, so that the smoke hid the light of day. At this time they stripped the hair from his skin, and anointed him with a sticky earth on which they placed gold dust so that he was completely covered with this metal. They placed him on the raft . . . and at his feet they placed a great heap of gold and emeralds for him to offer to his god. In the raft with him went four principal subject chiefs, decked in plumes, crowns, bracelets, pendants and earrings, all of gold. They, too, were naked, and each one carried his offering. . . . When the raft reached the centre of the lagoon, they raised a banner as a signal for silence. The gilded Indian then . . . [threw] out all the pile of gold into the middle of the lake, and the chiefs who had accompanied him did the same on their own accounts. . . . After this they lowered the flag, which had remained up during the whole time of offering, and, as the raft moved towards the shore, the shouting began again, with pipes, flutes, and large teams of singers and dancers. With this ceremony the new ruler was received, and was recognized as lord and king.

Just as Ponce de Léon became obsessed with the Fountain of Youth, Gonzalo Pizarro's Holy Grail became the secret lake of gold, and he thought he knew exactly where to find it. He selected for his second in command young Captain Orellana, his nephew from Trujillo, and assigned him to recruit men and gather horses, and rendezvous in the valley of Zumaco at the headwaters of the Coca River.

A second objective of the expedition concerned something even more valuable to Spain than a lake of gold. For 4000 years, Arab traders had been carrying the distilled bark of a small evergreen of the Laurel family — cinnamon — from Ceylon and Sumatra, sailing across the Indian Ocean or traveling by camel caravan overland to Egypt. As cinnamon's taste and medicinal qualities gained popularity, it was traded with Macedonia and Phoenicia. It was brought by Venetian sailors to

Italy; as its value grew, Venice and the other Italian trading cities established a monopoly of the spice trade in Europe and amassed wealth enough to build palaces of marble and gold. The disruption of this trade by the rise of the Mamluk Sultans and the Ottoman Empire shifted the flow of wealth to Constantinople and compelled the Europeans to venture out in search of new routes to the Indies, perhaps even by sailing west.

At the start of the 16th century, the Portuguese had discovered Ceylon, and by 1518 they had established a commanding military presence there. By the time that the Pizarro brothers left Spain, Portugal had an effective monopoly on the world's supply of cinnamon; if unchecked, Lisbon was on track to become the next Constantinople (although its plans would be derailed by the Dutch East India Company's capture of Ceylon's plantations in 1640).

With two very good reasons to mount an expedition into the South American interior — gold and cinnamon — all Gonzalo Pizarro needed were some skilled and adventurous junior officers.

Francisco de Orellana had by this time mastered several native languages, and with his recruitment skill the El Dorado expedition disembarked with 240 Spaniards and 4000 Native Americans. Crossing the Andes was a logistical challenge for such a large party, and Gonzalo Pizarro's brutality was of no use against hunger and thirst. After following the courses of the Coca and Napo rivers, the expedition started running out of provisions in mid-summer. About 140 of 220 Spaniards and 3000 of the 4000 natives died. In December 1541 or January 1542, Pizarro decided to turn back, but first dispatched Captain Orellana and 57 men to continue sailing downriver in search of food, with instructions to catch up to the General's party should any be discovered.

Among the 57 men sent with Orellana was Father Gaspar de Carvajal. Two hundred fifty years later, Lewis and Clark would become for North America what Orellana and Carvajal were about to become for South America. Father Gaspar's diary is the source of what we know of the failed expedition for El Dorado and the Land of Cinnamon, and as it turns out, the earliest recorded Western observation of the sustainable agriculture of the New World.

4

El Dorado

Putting into the middle course of the Coca River above present-day Santa Rosa and travelling downstream out of the Andes through a series of rapids and waterfall portages, Orellana's two hand-crafted boats, *San Pedro* and *Victoria*, each with 24 oarsmen, encountered a much larger and fast-flowing river. Carvajal called it the Marañón, but later it came to be called first the Orellana and then by the name we know it today, the Amazon.

Maps of the Amazon preceded Orellana's 1542 discovery by at least a few decades. It appears as a coastal river, unnamed, on maps and globes by Waldseemüller (Universalis Cosmographia, 1507) and Schöner (1515 and 1520), and it seems likely that it would have been mapped at least to the Rio Negro by the cartographers accompanying the great circumnavigation voyages of Zheng He in 1421–1424.[1] Vicente Yáñez Pinzón was the first European explorer to enter the river from the east, in 1515. Orellana appears to be the first foreigner, however, to enter it from the west, from the Andean plateau, and to navigate all the way to its mouth on the Atlantic coast.

As Pizarro may have expected when he released the detachment, they did not find food. When hunger reduced the party to eating their own shoes, Father Gaspar described their desperation:

We came to such great necessity that we ate not only the leathers, but the thread and soles of our shoes, cooked with some grass. So great was our leanness that we could not stand, but walked on all fours, leaned together to stand and walk a few paces, or made pilgrim's staffs so we could go up into the hills to look for some roots to eat. Some of the companions ate some unknown grass, and soon were on the verge of death, as crazy as if they did not have a brain.[2]

Eventually, in early January, they reached the outskirts of settled areas. Since they were in no condition to fight, they avoided contact whenever possible, looking for an opportunity to sneak into a home or village and steal some food. Failing that opportunity, they finally resolved to gather on the shore, load their arquebuses with powder and shot, and march in formation toward a small town. According to Carvajal:

> The spirit that all received in seeing the town made them forget all past fatigue, and the Indians, seeing their show of force, fled the town, leaving behind all the food.

The men fell into disorder as they gorged themselves, but Orellana cautioned them not to drop their guard, so they kept swords under their armpits as they gathered up what they could carry and hastened to get back to the safety of mid-river. As they returned to their boats, however, they saw there were natives observing them, so Orellana approached and, using the native languages he knew, told them there was nothing to fear. When two ventured out onto the beach, he gave them gifts and flattered them. They called forward a leader, who, perceiving Orellana's goodwill, embraced him. Orellana brought out gifts and gave the leader a dress jacket. The leader asked Orellana what he wanted, and Orellana said he wanted only food for his companions. Father Gaspar says:

> Soon the Cacique sent orders that the Indians bring food, and with very great haste they brought in abundance what was necessary; meats, partridges, turkey hens, and fish of many types. And later, the Captain thanked the Cacique and said to him that the gifts were not for himself, but went with God.

Having made friends, and feeling much safer, the party decided that they had need of a larger boat, one with sails, if they were to stay out of harm's way as they continued downriver. The one thing they needed most to build such a boat was nails. There were iron ores in the area, so Orellana directed the construction of a furnace and smelter, and asked the cacique for help in making charcoal. Two of the companions, Juan de Alcántara and Sebastián Rodríguez, were expert at making charcoal. Whether the culture which Orellana's party had encountered had their own experience with charcoal-making is not revealed.

The soldiers were still very malnourished and easily fatigued, but they set about mining and smelting the iron and building an ironworks. Father Gaspar explains:

> They made bellows of half boots, and in a similar way, made all the other tools, and (with direction from Alcántara and Rodríguez) harvested a good amount of coal. This required going to the hills to cut firewood and to bring it back to the town, and also going some leagues to find iron, and dig it from the ground, and this was all a very great effort. As they were skinny and unskilled at these tasks, they could often not manage the loads, or have strength to cut wood. But they took turns, and the weakest sounded the bellows and others carried water, and the Captain worked beside his men, in everything, speaking encouragement, so that all shared the understanding of what was needed and why. Our company built the factory, and despite our weakness, it occurred in such a good way that it impressed the town, and in twenty days, by the Grace of God, we made two thousand very good nails and various fittings, and built the brig.

As they continued downriver, the settlements grew larger and more impressive, and the inhabitants were not always friendly. Carvajal describes near-continuous waterfront settlements that stretched more than 80 leagues (240 miles[3]), with many larger boats docked there. As they traveled, they kept to the center of the river, secured their boats together when defense was needed, and kept unwelcome war parties at bay with shots from arquebuses or quills from crossbows.

On May 12, they arrived at the province of a chief named Machiparo who lived in a great glimmering white city that the expedition had seen from six miles upriver. They were told that Machiparo's army was

50,000 strong. When Orellana attempted to bypass the principal port and continue downriver, he was intercepted by a

> very great amount of canoes, as high as a man, made ready for war, and covered with shells of lizards and leathers of manatees and deer. They made a very great outcry, beating many drums and sounding trumpets of wood, threatening that they had come out to eat us.

Orellana ordered the boats into defensive position and warning shots fired, but the powder was too damp to ignite. There then ensued a great battle. The conquistadors fired their crossbows, which halted the advance of the defense forces, but those on shore took to the water and dragged the boats close enough for hand-to-hand fighting. Orellana directed a lieutenant to charge the town and put it to the torch. With sword, pike, and crossbow, the lieutenant led 25 men in a frontal assault, and the townspeople fled. The fight at the boats halted, and the attackers withdrew. Going into the town, Carvajal recounted:

> There was great amount of food, turtles in corrals and cisterns of water, and much meat and fish and sponge cake. There was so much abundance to eat it could have fed a thousand men for a year. Having seen to the security of the good port, the Captain decided to gather food and to rest, and so ordered Cristóbal Maldonado to secure what they needed.

After they had gathered more than 1000 turtles to take with them, their attackers returned. Orellana ordered a retreat and left Maldonado's squadron of 10 men to fight a rear-guard action against more than 2000 well-armed natives. Wounded in the arm and face, Maldonado kept up the defense until the boats were launched, and then, as he attempted to withdraw, was overtaken by some 500 attackers, who slew four of the wounded. Orellana, watching this, ordered his troops from the boats, and they charged into the melee, rescuing Maldonado's remaining force and securing the retreat.

There followed days and weeks of skirmishes with pursuing natives. At times the company would find an island where they could camp, repair the boats, and cook food, but even there they would have to fend off attacks by canoe at all hours of the day or night.

Farther downriver, the cities grew still larger and better defended.

There walked between the people and their war-canoes four or five wizards, all whitewashed and their mouths ringed with ash, that threw ashes to the air, or held in their hands rattles and censers, with which they walked, throwing water by the river, in the way of spells, and after they had given an incantation to our brigs in this way, called incitements to the warriors, and then came bugles, trumpets and drums and a very great outcry of voices shouting.

As they put backs to oars and fled, they were followed and harassed for two days and nights, during which none of them could sleep, all the while passing densely populated riverbanks. Father Gaspar recorded that the people all spoke the same language, and for more than 80 leagues, were in such density of settlement

that a crossbow shot could connect town to town and the farthest separation less than an average league, and at least five cities lasted entire leagues without separation from house to house. That was a wonderful thing to see: even though we went past fleeing and could not examine the place or know what was there, or anything of the peoples in those regions, according to the disposition of what we saw, the interior must be populated much as what we had seen, and thus, this area ... we must say is "grandísimo."[4]

There were many roads here that entered into the interior of the land, very fine highways... Inland from the river at a distance of 2 leagues [6 miles] more or less, there could be seen some very large cities that glistened in white. This, the land, is as good, as fertile, and as normal in appearance as our Spain, for we entered it on Saint John's day and already the Indians were beginning to burn over their fields.

It is a temperate land, where much wheat may be harvested and all kinds of fruit trees may be grown. Besides this, it is suitable for the breeding of all sorts of livestock, because on it there are many kinds of grass, just as in our Spain, such as wild marjoram and thistles of a colored sort and scored, and many other very good herbs. The woods of this country are groves of ever-green oaks and plantations of cork trees bearing acorns (for we ourselves saw them) and groves of hard oak. The land is high and makes rolling savannahs, the grass not higher than up to the knees, and there is a great deal of game of all sorts.[5]

Finally, some 60 miles downriver from the dense settlements and growing hungry and weak again, they resolved to make an attack on what appeared to be an undefended town, take some food, and leave quickly. The local residents, seeing them approach the port, withdrew to defend the central warehouse, but with shots from their arquebuses, the Spaniards drove them away. They discovered the town was far wealthier and more developed than they had anticipated, and so they remained there three days. Carvajal reported:

In this town there is a very great amount of very good sponge cake, that the Indians do of maize and ayuca, and much fruit of all sorts.

Orellana's first glimpse of the dark earths may have come in an unexpected way. After eating enough to recover, his men marched some distance by the road, and on June 3, 1542, they arrived at a large tributary flowing into the Amazon. The water flowing from the river was black and thick "like blood" and resisted mixing with the other rivers. They called it the Rio Negro. This "Black River" is not colored by terra preta soils, but rather by the leaves of trees that fall into its tributaries and are washed down into its warm waters to brew like a reddish-black tea. Carvajal was neither a soil scientist nor a naturalist, and he made no entry regarding black earths. However, high on the bluffs that line the Rio Negro can be seen terra preta terraces, and occasionally, in cuts and slices calved by erosion in the rainy season, are revealed across the face of the bluffs the deep topsoils that supported the great lost cities of the Amazon, held to the banks by the thick roots of trees.

5

The Great White Way

WHERE ORELLANA'S PARTY WAS AT THAT MOMENT — at the confluence of the Negro and Solimões Rivers — is today the Brazilian city of Manaus, constructed to house the 19th-century opulence of slave lords and rubber barons. Today the city is being refurbished for the 2014 World Cup soccer games, fighting back the decay and entropy of the tropics that, employing merely mildew, moss, and a few aggressive vines, constantly reclaim cement and iron, tile and tin. Saran wrap, styrofoam, coconut husks, and beer cans wash from the streets of Manaus and glide into the river. Displaced Indian refugees hover around campfires in city lots and barrio landfills.

Father Gaspar described what Orellana's expedition saw in 1542:

In this town were houses of pleasing interiors with much stoneware of diverse forms. There were enormous pitchers and vases, and many other smaller containers, plates, silverware, and candlesticks. This stoneware is of the best quality that has ever been seen in the world, and even that of Malaga does not equal it. It is all enameled with glass, of all colors and the brightest hues. Some are drawn to frighten, but on others the drawings and paintings are delicate depictions of nature. They craft and they draw everything like the Romans. There were ornaments of gold and silver, and in this house were two idols woven of feathers of intricate design, and designed to frighten. There were giant statuary and in one there were working arms and knees, run by gears and wheels. The statues' heads had very great ears, with ornate earrings. And also in this town were much gold and silver; but our intention was not to look for wealth but to eat and to try to discover how we might save our lives.[1]

From this village there went out many roads, fine highways to the inland country. The Captain wished to find out where they led to. For this purpose he took with him Cristóbal Maldonado and the Lieutenant and some other companions and started to follow the roads. He had not gone half a league when the roads became more like royal highways, and wider. When the Captain perceived this, he decided to turn back, because he saw it was not prudent to go any farther.[2]

A few days later, Carvajal described making port at a medium-sized village and being astonished by its feats of architecture.

In this village was a very large public square, and in the center of the square . . . were two towers, very tall and having windows, and each tower had a door, the two facing each other, and at each door were two columns; and this entire structure that I am telling about rested upon two very fierce lions, which turned their glances backward as though suspicious of each other, holding between their forepaws and claws the entire structure.[3]

Back on the river, the routine continued. The travelers bypassed large population centers, occasionally got into skirmishes with armies numbered in the thousands, and looked for small towns into which they could land a raiding party, steal whatever food they could lay their hands on quickly, and resume their journey downriver. Father Gaspar observed and recorded fine colored clothes of cotton and wool, wealthy port cities, ceremonial buildings with roofs clad in macaw feathers, woven tapestries that narrated historical events, elaborate carved furnishings, and even machinery. On June 25, Father Gaspar wrote:

Among some islands which we thought uninhabited, but after we got to be in among them, so numerous were the settlements which came into sight . . . that we grieved . . . and, when they saw us, there came out to meet us on the river over 200 pirogues, that each one carries 20 or 30 Indians and some 40 . . . colorfully decorated with various emblems, and they had with them many trumpets and drums . . . and on land a marvelous thing to see were the squadron formations that were in the villages, all playing instruments and dancing about, manifesting great joy upon seeing that we were passing beyond their villages.

Towards the end of June, Orellana's men had an encounter with natives from whom the river now takes its name. Rounding a bend, they came upon a group of well-armed, armored warriors lying in wait. Near the ground chosen by their attackers, the river suddenly became shallow, and when the Spaniards jumped out into the chest-high water to free their boats from the mud, the warriors attacked, engaging them in close fighting. Gaspar observed that the attackers were led by 10 or 12 "captains" who fought spiritedly at the front. As they advanced, the Spaniards could see that these captains were women. Father Gaspar described them this way:

These women are very white and tall, and have very long hair, bound and shaken wildly at the top, and are very bold and walk naked, but for leathers over their shames. They carry scimitar knives and bows and even if you shoot them with arrows in their arms, they still fight as much as ten men; and these women put so many arrows into one of our brigs, that it looked like a porcupine.

Orellana's expedition continued another two months, intermittently raiding towns for food and defending from attacks, until finally, in August 1542, they reached the Atlantic. This is the same year that João Rodrigues Cabrilho landed in San Diego Bay, California; Hernando de Soto died of fever, his expedition in ruins after innumerable atrocities inflicted on the indigenous populations of Florida, Georgia, North and South Carolina, Tennessee, Alabama, Mississippi, Arkansas, and Texas; and a Portuguese ship, blown off its China course, landed in the Japans, making European contact with Tokugawa Ieyasu's Edo Empire.

With neither sails nor compass, and with no safe harbor where they were, Orellana pushed his surviving soldiers to row north around the coast to Paria, in Venezuela. It took seven days and nights of rowing just to escape the tidal estuary and get out to sea, but they were fortunate to have fair weather for the next few weeks. On September 11, they reached Nueva Cádiz, on the island of Cubagua, and from there took passage on a Spanish brig to Santo Domingo, Venezuela. From there, Orellana and Carvajal returned to Spain to inform the King of the discovery of the river.

In 1542, Carvajal's journal was excerpted for use in *General and Natural History of the Indians,* by Gonzalo Fernández de Oviedo. Oviedo's book was not published, though, until 1855. The excerpts then lent fuel to the myth of El Dorado, the Lost City of Gold. They also reignited the popular legends of Amazon warriors and, later, *Wonder Woman, Xena: Warrior Princess,* and *Tarzan and the Amazons.*

Selections from Carvajal's journal were translated by the Chilean scholar José Toribio Medina and printed in 1895 as *Relación* by Medina's family in Spain; it has never been published in English translation. The quotes used here alternate between my own translation of the Spanish original and Medina's version.[4]

Back in Spain, Orellana married and then returned with his wife to the New World. He also brought along an expedition of 200 foot soldiers and 100 cavalry commissioned by the King of Spain to explore and colonize "New Andalusia."

Landing at the mouth of the Amazon at Christmas 1545, Orellana constructed two great riverboats and sailed 300 miles up the river until he ran low on food and was forced to divide his party to search for provisions. Traveling upriver was a different proposition from coming down; Orellana lost the main channel, and, while searching to regain it, was attacked by well-prepared warriors. Seventeen of his party died from poison-tipped arrows. Orellana himself was one of the casualties, succumbing to the effects of his wound in November 1546, off-course, lost, and starving.

With Orellana's death, his story fell into the realm of myth and fable. The wealthy complex societies of the central Amazon recorded by Gaspar de Carvajal were thought by European scholars to be as fantastic as El Dorado or the Amazon women warriors. Later, even archaeologists disputed the notion that a large, complicated Amerindian culture could have arisen in the equatorial tropics of Brazil, where soil is notoriously too poor to grow anything more than sweet but unnourishing fruits and jungle spices.[5]

The intrepid explorers of the 19th and 20th centuries who hacked through the jungles of Mesoamerica and South America beyond the Amazon did find amazing things — mountainous Inca citadels such as Machu Picchu, the pyramid cities of the Itzá Maya — and remarkable feats of engineering in such places as Cuzco, Lake Titicaca, Palenque, and Teotihuacán. Yet, in all the sprawling Amazon itself, until just the past decade, no expedition found anything to confirm the Dominican priest's account. There were no glistening white cities, no incomparable stoneware and tapestries, no machines, no wide causeways extending six miles inland. Just vines, grasses, ferns, massive trees, mud, and mosquitoes.

What modern explorers could not see, for all the jungle, was the black gold that lay beneath that vegetation.[6] Without seeing that, they could not explain the existence of the huge populations witnessed by Orellana.

And when the Amazonians went extinct from epidemics of the Old World — diseases that no amount of military training and discipline could prepare their sons or daughters for — they took with them not only the secret of El Dorado, but the recipe.

6

The View from the Bluff

IT HAD BEEN FIVE GENERATIONS since the inhabitants of Açutuba, Brazil, had seen such a strange vessel pass by in the broad black river below their village. In the year 1423 (counting by a Gregorian calendar they would not have known), their great-grandparents may have witnessed the passage of one or more Chinese vessels that explored the Negro and Solimões Rivers, turning back only when the rivers grew too narrow for their ocean-going crafts to continue.[1]

Now, in early June 1542, the Açutubans once more watched, below them in that river, a curious wooden sailing vessel of foreign design pass under their high bluff, manned by soldiers in metal helmets and breastplates. The passage of the strange vessel caused some discussion in the days and weeks that followed, but, seeing no more of its kind, the memory passed into the village's history, and life went on as it always had.

On this hillside overlooking the Rio Negro, about 30 miles upstream from present-day Manaus, there had been continuous settlements for nearly 2000 years. Daily life had changed very little over that time, although military empires and dynasties came and went, religious revivals waxed and waned, and some years were better than others for fishing and hunting.

Food was seldom a problem, although dietary preferences and available foods evolved. In the earliest settlement period, from 450 BCE to 400 CE, people hunted the jungles, fished the river and its tributaries, and gathered fruit, nuts, roots, drugs, and fibers from the forest. The abundance of the natural environment surrounding them, in all seasons, meant there was seldom hunger. The exceptions were in those years when Açutuba suffered the intervention of distant lords or marauding armies.

During their earliest stage of village development, the people of Açutuba filled holes in the ground — some created by tree falls in storms, others by burrowing animals — with the refuse from their kitchens. This refuse included fish and animal bones, broken bits of pottery, fruit skins, nut husks, turtle and oyster shells, and the ash and cinders from their fires. The places where they disposed of their wastes also were used as places to defecate, and so tended to be located at some distance downwind from their homes.

Over time, these rubbish sites became extremely rich in phosphorus, nitrogen, potassium, and carbon. Later, when the villagers acquired seeds for corn, squash, gourds, beans, chilis, palms, and fibers such as cotton, they dug into these aged rubbish piles and spread the rich black humus on their fields, or, if conditions suited, planted crops directly into the decomposed mounds. This period of early agriculture, as opposed to the earlier agroforestry, is called the Manacapuru, and lasted from around 400 to 900 CE.[2]

Towards the end of the first millennium, the region went through political turmoil, and the Manacapuru culture gave way to the Paredão and then Guarita cultures, a change that was characterized by a distinctive shift in the pattern of the ceramics, to what archaeologists call the Amazonian Polychrome Tradition. Cultivation moved farther away from a pattern of landscape domestication[3] employing guilds of native food plants in biomes that resembled natural ecologies, and more into a pattern of crop rotation within and surrounding the village. Defensive necessity may have influenced this shift, but it was fortuitous, because in transporting the rich black soils from their sites of origin to more distant fields (creating the terra mulata soils often found around terra preta sites), the post-Manacapuru peoples came to better appreciate the need to continuously renew their reserves of dark earths.

Archaeological work at the Açutuba site tells us that the recipe for terra preta was well understood by the time Orellana passed near this village. It is possible that the Açutubans had a kiln that they built and operated for making their charcoal, and they may have used the larger pieces in smokeless stoves inside their homes, and heaped the finer powders onto their compost piles. By doing this simple exercise, they would have ensured that the land they bequeathed to their successors would always be richer than the land their ancestors had bequeathed to them.

Elsewhere in Brazil, at a site called Santa Catarina, anthropologist Lilian Rebellato uncovered a mound of oyster shells overlain by a dark band of rich terra preta. Below the oysters were a more random series of strata showing haphazard development of the black earths. Above the oyster shells was the methodical production of terra preta for agriculture. Rebellato theorized that this shift illustrated the ancient culture's "Aha!" moment.

While communication between villages was extensive, it was not universal, and development of the terra-preta-making culture occurred at different times in different places. The oldest field discoveries have carbon-dated some Brazilian Amazonian terra preta sites to 8500 years ago.[4] Elsewhere, such as in the Colombian Amazon, there are terra preta sites that date back to only 384 CE.[5] From this it is possible to conclude that adoption of the practices of terra preta soil-building, while widespread, disseminated very slowly from region to region.

Sadly, it was not long after the passage of the strange boat filled with ironclad men that the Amazonian tradition of generational gifting disappeared — vanished from the face of the Earth as if it had never existed.

Estimates of the dead — from viruses and bacterial infections brought from Europe (and conceivably from Asia), to which the natives of the Americas had no immunity, and from the cruelty that came from the conquest — range from 90 to more than 99 percent.[6] Controversy so envelops estimates of the pre-Columbian population of the Americas that not even the order of magnitude is known. The Americas could have held as many as 1 billion, or as few as 30 million.

Any smaller estimates than those are absurd on their face, as shown by recent archeological work in Hispañola. We now know with a high degree of certainty that the population of that island alone, prior to the arrival of Columbus in the 15th century, was about 18 million. Hispañola has a land area of 29,529 square miles (76,480 km²). North and South

America have a land area of 16,428,261 square miles (42,549,000 km²). If, for sake of argument, we say that only a tenth of that total land area was suitable for cultures of the type known to exist in Hispañola, then the pre-contact population of the Americas could have reached 1 billion. Plummeting to 500,000 within 200 years after contact, the extinction rate would have been 99.95 percent.

Central Mexico, the Andes, and some Caribbean islands were likely to be among the most densely settled regions, but Amazonia, with its abundant food supply and favorable climate, could also have been a region of high population, comparable to Central Asia in the same period.[7] Estimates range from 1.5 million[8] to 30 million[9] people in the Amazonia of Orellana's time,[10] even higher than the present population. If the high estimate is true, the extinction rate, just in the Amazon, was 99.3 percent.[11]

How could such a severe die-off have occurred? Even Europe's Black Death, in its worst outbreak from 1347 to 1351, killed only a third of those who were infected.

Several factors were to blame. First, while Europeans were acclimated to the diseases of domestic animals such as swine and avian flus, rodent- and flea-borne poxes, and bovine rinderpest (related to measles), Native Americans lacked those immune antibodies. While they had domesticated many indigenous animals, they had no experience with swine flu.

Hernando de Soto's expedition to the Mississippi and Tennessee River Valleys in 1539, like Orellana's, reported impressive cities — and it brought with it 600 soldiers, 200 horses, and 300 pigs. Swine are optimal hosts for mutations of anthrax, brucellosis, leptospirosis, taeniasis, trichinosis, and tuberculosis. Escaping to the swamps and forests of North America, de Soto's pigs bred quickly and may have transmitted pathogens.

The second factor to blame for the great American die-off is that there was no indigenous knowledge of vector-borne disease or the proper response. Europeans knew enough to quarantine victims, kill rats, burn blankets, and evacuate cities to the countryside. In the Americas, friends and family gathered with shamanic healers at the sufferer's bedside to wait out the illness. Blankets and the clothes of the deceased were passed to relatives and neighbors.

Thirdly, from studies performed by Francis L. Black, an epidemiologist at Yale University, we know that Native Americans have unusually homogeneous immune types. The major histocompatibility complex (MHC) is a large genomic region found in most vertebrates. It plays an important role in the immune system and autoimmunity — the way populations, including humans, respond to disease attacks — because it is this genomic region that distinguishes cells as self or non-self.

Pathogenic bacteria and viruses are constantly mutating and seeking new hosts. They may find a suitable host in an individual of one MHC type, only to be killed by the MHC response of another. For most animals, there is safety in numbers, because the more animals, the more different immune types there are, and epidemics die out quickly.

Black proved that Native Americans have extremely similar MHC types. For one person in three, they are virtually identical. Among Africans, the corresponding figure is one in 200. While contact with smallpox, typhoid, bubonic plague, influenza, mumps, measles, whooping cough,

Selecting for Annihilation?

In 1995, Swiss biologist Claus Wedekind discovered MHC-dissimilar mate selection tendencies in people of European decent. In Wedekind's experiment, a group of female college students smelled T-shirts that had been worn by male students for two nights, without deodorant, cologne, or scented soaps. An overwhelming number of the women preferred the odors of men with MHCs that were dissimilar to their own. However, their preference was reversed if they were taking oral contraceptives that apparently can interfere with MHC signals. A different study in 2005, on 58 test subjects in Southern Brazil, contradicted Wedekind's findings. While not conclusive, these studies are indicative that the human MHC genome in the Americas may have differed from the same genome in Europe, and that prior to contact, South American populations did not select mates for MHC dissimilarity and so set themselves up for annihilation.

We know from recent mitochondrial DNA studies that it was a very small group of individuals, possibly as few as 80, who survived the Arctic crossing from Asia and went on to populate the Western Hemisphere at the end of the most recent Ice Age.[12] Is it possible that this small group carried the MHC genetic anomaly that later erased 15,000 years of cultural history?

cholera, diptheria, malaria, scarlet fever, yellow fever, syphilis, chlamydia, gonorrhea, and herpes did not decimate the population of Africa, the same cannot be said of the Americas. Once a virus found a favorable host that lacked immunity, chances were good that the next host would have the same immune deficiency. The Americas were every virus's free buffet.

After the passing of the plagues, and the brutal conquest that ensued, those Native Americans who remained were reduced to a state of extreme poverty. Before European contact, the people of Amazonia cultivated or managed at least 138 of the 257 plant species cultivated in the Western Hemisphere. Two centuries later, they were reduced to farming only a handful.[13]

Anthropologist Charles C. Mann says, "The pall of sorrow that engulfed the hemisphere was immeasurable. Languages, prayers, hopes, habits, and dreams — entire ways of life hissed away like steam."

When the civilizations of the Americas perished, with them perished the agricultural sciences gained from millennia of field trials. Countless valuable domestic cultivars, unable to self-propagate, went extinct. Where great shining cities had stood, vines and moss covered the façades; trees broke through the paving stones and engulfed buildings. Rain rotted away the roof timbers, and insects ate the parchment of scientific and literary manuscripts, leaving but a few to the bonfires of the conquerors.

So great was the burst of vegetation over open fields and mounded cities in the Western Hemisphere that the carbon drawn from the air to feed this greening upset atmospheric chemistry. Analysis of the soils and lake sediments at the sites of both pre-contact population centers and sparsely populated surrounding regions reveals that the reforestation of land following the collapse drew so much carbon out of the atmosphere so rapidly that Europe literally froze.[14]

That period of global cooling, which was most intense from approximately 1500 to 1750, is known as the Little Ice Age. The Little Ice Age ended the Medieval Warming Period that corresponded to the time when the Amazon was not jungle, but tall white cities with many miles of wharf-front, wide causeways extending inland, and highly cultivated societies.

Confederados

A T THE END OF THE AMERICAN CIVIL WAR, in 1865, living condi-tions in the Southeastern United States became intolerable for a number of planters, a class of gentleman farmers produced by the antebellum era.

Before the war, many hereditary Southern fiefdoms had been made wealthy by the institution of slavery, an economic artifice that allowed a master class to wheel and deal in cotton futures or cavort about the capitals of Europe in search of fashions and fancies, while out in the fields in the sweltering summer sun horsewhipped Africans picked thorny bushes clean of cotton fluff with callused and bleeding fingers.

After the war, the African slaves were emancipated legally, but were kept economically and politically enslaved. The masters of the new sys-tem were the bankers and corrupt governors, Northern "carpetbaggers," who made the real money. Plantation owners became low-paid middle-men and even lost the right to vote if they had previously served in the Confederate Army or ever voted for a secessionist legislator.

Rather than suffer these indignities, many planters decided to sim-ply pack up and leave. The Confederate exodus was the largest emigrant movement in US history, rivaled only by "back to Africa" campaigns. One wave went to Mexico, wishing to remain close "if an opportunity to invade the United States arose." Another joined the Egyptian army, not for the purpose of colonization, but for "vindication, for adventure, and for wealth."

Confederate Admiral John Tucker left the US and became a rear admiral for the combined Peru-Chile fleet in the war against Spain. Tucker took other ex-Confederates with him to serve in the fleet; that group later surveyed the Amazon River.[1]

Confederates also settled British Honduras (now Belize), Venezuela, and Argentina, but it was Brazil that won the greatest number, and it wasn't just the scuttlebutt from Tucker's surveyors that drew them. Brazil offered cheap ship passage, temporary housing, mortgage loans, sale of any land without restriction, promise of rail access for the more distant settlements, full citizenship after two years of residency, and the *pièce de résistance*, legal slavery. The Brazilian government made it clear to prospective immigrants that if they came South, they could resume their Southern lifestyle.

A letter from a Brazilian émigré in the *Mobile Daily Register,* November 26, 1868, claimed that neither California nor Texas "could surpass the superiority of Brazil when it came down to climate, soil fertility, and individual rights."

In *Brazil, the Home for Southerners: or a Practical Account of what the Author, and Others, who Visited that Country, For the Same Objects, Saw and Did While in that Empire* (1866), the Reverend Ballard Dunn described his visit to Taperinha, one of the new Confederado settlements along the Rio Tapajós.

There are lines of swampy forest, and strips of arums, and clumps of bushes, all running parallel to the channel: seams left by the Amazons in sewing this patch-work together. Back of us the great cane field stretches half a mile or more in every direction, fresh, green, waving — the prettiest sight a planter's eyes could find. The cane is cut by hand, and brought to the brow of the hill on ox-carts; there it is thrown into a long shoot, which deposits it cleverly in the mill-house. No wonder that the cane thrives here; the ground is a rich black loam, two feet thick; we see it in the road-cuttings, and it spreads away beyond the field far into the thick forest.

It is curious to note what gave this land its richness. The refuse of a thousand kitchens for maybe a thousand years, together with the numberless palm-thatches, which were left to rot on the ground as they were replaced by new ones. For the bluffs were covered with Indian houses, "so close together," says Acuña, "that from one village you can hear the workmen of another." The people made coarse pottery and marked it with quaint devices. We find fragments scattered everywhere, and for years Mr. Rhome has been making archaeological collections, including all sorts of curious things: a whistle, vultures' heads, frogs, and a cock with comb and wattles complete. The Indians were cremationists: burning their dead and burying them in jars under their floors; and several of these burial urns have been obtained at Taperinha. Stone implements are not common: a few handsome axes and arrowheads were picked up here, and below the hill.

Generally this black soil does not extend more than half a mile from the face of the bluff; after that the land is red sandy clay, for mold does not form in the forest as it does at the North: the leaves fall singly and are never packed together by a blanket of snow.[2]

When geologist, clergyman, and explorer James Orton visited Confederado settlements on the Belterra plateau near Santarém, Brazil, in 1868, he reported that "the soil is very black and fertile. It beats South Carolina, yielding without culture thirty bushels of rice per acre." The choice of the sites was no accident, but had come from extensive surveys of the soils between Santarém and Manaus by Confederado Landsford Hastings in 1866.[3]

An Austrian geological exploration led by Friederick Katzer in 1895–98 looked at these soils in greater detail and gushed that the region's "more distinguished wealth lies in its soil." Katzer estimated that the "Schwartze Erde" extended over more than 50,000 hectares immediately south of Santarém.[4]

On his return from Brazil in 1898, Katzer was appointed director of the Geological Institute in Sarajevo. He authored more than 140 scientific works, including chemical studies of his samples of terra preta from Santarém, but almost all of Katzer's collection was destroyed, along with the National Museum of Bosnia-Herzegovina, during the Bosnian conflict of the 1990s.

8

Hartt's Breakthrough

G ROWING UP IN WOLFVILLE, Nova Scotia, in the mid-19th century, Charles Frederick Hartt showed an unusual aptitude for languages. By the time he went to Acadia University, he could read several romance languages and was fluent in Portuguese, picked up from a neighborhood shoemaker. In the 20 years to follow, he would master more than 10 languages, including several indigenous native dialects of the Amazon.

Hartt's studies in geology, together with his skill in Portuguese, won him, at age 15, a scholarship to the 1865 Thayer expedition to Brazil mounted by Louis Agassiz of Harvard University. Agassiz was attempting to refute Darwin by finding evidence of glaciation at the equator. His notion was that if natural forces destroyed all land mammals, say 4000 years ago, divine intervention would have been required in order to bring about such diverse speciation in the intervening amount of time, and hence creation fit within Biblical time frames.

Agassiz was so impressed with Hartt that on their return from Brazil he placed the young man's name forward to be the inaugural Professor of Geology at Cornell University. Hartt took the position, and then, at a mere 30 years old, announced publicly that his benefactor, Agassiz, was clearly wrong about Darwin, and what was being claimed as evidence of tropical glaciation was merely bedrock weathering.

Such open criticism did not endear him to many in the geological community, which held Agassiz in high regard, nor did it help the prestige of Cornell in its early years, but it did win Hartt some serious supporters, including Colonel Edwin P. Morgan, who offered to fund two more Hartt expeditions to South America.

By the time he launched the Morgan expeditions of 1870 and 1871, Charles Hartt had already returned to Brazil in mid-1867 to explore the coastal morphology and coral reefs. The Morgan support allowed him to explore the lower Amazon with his Cornell students. Then, in 1874, the university granted him leave to conduct a complete geological survey of Brazil.

Funded by the Brazilian government, which was hoping to get treasure maps of exploitable minerals, Hartt and his Cornell students collected over 500,000 samples and made thousands of charts, but unfortunately worked too slowly and methodically for the patience of the Imperial government. Brazil is a large country, and the goal Hartt had set for his small work group, to produce a geological survey of its entirety, was perhaps overly ambitious, given the transportation of that era, the size of Brazil, and the difficult terrain. In 1878, Brazil's emperor cancelled Hartt's funding, and he was in Rio de Janeiro pleading to have it restored when he contracted yellow fever and expired before reaching his 38th birthday.

The sad part of the story is that during the time the project had been fully funded, Hartt had brought his wife and two young children to Rio de Janeiro; they had lived in a splendid apartment there where Hartt completed five books and more than 50 scientific studies, as well as designs, paintings, and etchings about the country. With the loss of support, he was forced to send his family home by tall ship, and then, less than two months later, he died, not in the care of his family, but in an austere rooming house in Rio.[1]

Hartt's part in the carbon story comes by the coincidence of his travels to Taperinha, the Confederado plantation mentioned by Ballard Dunn and soil sampled by Friederick Katzer. According to Eduardo Góes Neves of the Archeology Ethnology Museum of the University of São Paulo, "In [Hartt's] paper, *Contributions for the ethnology of the Amazonian Valley*, he suggested that the Taperinha village, a fluvial *sambaqui* [shell refuse heap] close to Santarém, must be very old because of its implantation into the landscape." Hartt and his Cornell students, presumably on the Morgan expedition with steamboats on loan from the government, encountered a group of Confederate expatriates. They were growing sugar cane at Taperinha "in plots of earth that were darker and more fertile than the surrounding soil."

Hartt immediately connected the dark earths to native settlement, using the term "kitchen middens" in an 1874 paper. His assistant, Herbert H. Smith, no stranger to Ballard Dunn's book, reported "the bluff-land owes its richness to the refuse of a thousand kitchens for a thousand years."[2] Smith would later write for *Scribners Magazine* in 1879: "The cane-field itself is a splendid sight; the stalks ten feet high in many places, and as big as one's wrist." The secret, he reported, was "the rich terra preta, 'black land,' the best on the Amazons. It is a fine, dark loam, a foot, and often two feet thick."[3]

One hundred years later, the US archaeologist Anna Roosevelt carbon-dated the shells collected by Hartt and confirmed them to have been of very ancient origin. She retraced Hartt's visit to Taperinha, and there, after re-digging the site, found some of the oldest ceramics on the continent embedded in the terra preta. The pieces of the puzzle, slowly assembling, began to fit together.

God may have made Darwin's finches, but the Amazon Indians made terra preta. Had he lived, Hartt might well have been the first to confirm that.

As it happened, the confirmation that terra preta was produced by ancient Amazonians came from the laboratory of Friedrich Katzer in Sarajevo in 1903. Katzer, a pioneer in the chemical analysis of soils, subjected his samples to loss-on-ignition, chemical leaching, and other tests that revealed an "intimate blending of mineral residuum, charred plant materials, and decomposed organics." Katzer concluded that these soils were not formed by nature, but were cultivated in ancient times when the region was more densely populated.

Sixty-three years would pass before Katzer's tests would be performed again, and his results confirmed, by Wim Sombroek.[4] During that hiatus, it was more or less taken for granted that the terra preta soils were the product of ancient human cultures that did not survive European contact. There was no interest in discovering the recipe, and perhaps even more curiously, no one seemed to want to know how it was that dense settlement patterns in the Amazon could produce such fertile, carbon-rich soils, while wherever else dense settlement patterns existed, soils were being mined and depleted.

In 1927, Henry Ford established his large rubber-growing community, Fordlândia, along the Tapajós river. In 1934, he relocated it upstream to the much more fertile setting at Belterra, 25 miles south of Santarém. Although Belterra has exceptional density of terra preta, centered on the indigenous town of that name, the Ford engineers apparently had no interest in that aspect of the site for growing rubber. They were only interested in the level terrain and deep-water port.[5]

When Walt Disney visited Belterra, he was smitten by the tidy, middle-American village in which Ford workers played golf on manicured links when not relaxing in air-conditioned buildings. Disney produced a short documentary entitled *The Amazon Awakens*.

When the Ford Company passed to Henry Ford II in 1945, Belterra was abandoned to the jungle. Today, the region is farmed by agroindustry for soybeans and biofuels, compacting the earth with giant tractors and combines, slicing, dicing, and sun-drying the microbial life out of the terra preta soils with giant turning plows, and leaching any remaining nutrients with ammonium fertilizers, pesticides, and herbicides, sending them running down to the river in the heavy rains.

9

City Z

I N THE 1970S, ANTHROPOLOGIST ANNETTE LAMING-EMPERAIRE
unearthed in eastern Brazil the skeleton of a 16-year-old, 59-inch-
tall woman whom she named "Luzia," a reference to the famous African
ancestor known as "Lucy." Laming-Emperaire died before she had a
chance to date her find, but 20 years later, Walter Neves found the skull
in a museum in Rio de Janeiro and carbon-dated it to between 10,500
and 9500 BCE. The find was eventually confirmed by the remains of
over 70 individuals with similar characteristics, uncovered in that same
region.

Using 86 complete genomes of maternally inherited mitochondrial
DNA (mtDNA), a group led by Nelson Fagundes determined that all
Native Americans are descended from a single founding population,
rather than being the product of multiple Asian migrations. The founding group, comprised of just 80 to 500 individuals, first reached the
New World 15,000 to 19,000 years ago. Those dates also correspond
to the opening of an ice-free corridor along the western coast of North
America, following a comet impact complete with megafauna extinc-
tions and dramatic climate effects.

Putting Fagundes's DNA tests with Neves's fossil dates, it is reasonable to suggest that the first members of the current human population
reached the Amazon sometime between 19,000 and 12,000 years ago.
Evolution is not so rapid as to have changed their physical attributes
much between then and now, and it is possible to imagine that the

hunting and gathering, plant and animal domestication, and highly adaptive lifestyles also changed very slowly in a region that has few natural catastrophes, a pleasant year-round climate, and abundant wild foods. The Amazon residents' biggest threat was probably one another.

Archaeologist Michael Heckenberger has spent so much time in the Kuhikugu village in the Xingu region of the upper Amazon that the Kuikuro have given him his own hut. The Kuikuro are likely descendants of the original inhabitants of the region.

Heckenberger's archaeological site is in an area that was known for its fierce tribes of cannibals. It was also known for the lost expedition of British explorer Percy Harrison Fawcett, vanished in 1925, looking for what Fawcett called "City Z."[1]

In 2005, *The New Yorker* staff writer David Grann visited the Kalapalo tribe in the Xingu in search of traces of Fawcett's expedition. He discovered that the Kalapalo had an oral history of a monumental civilization that may actually have existed near where Fawcett was looking. Grann's findings are further described in his book *The Lost City of Z: A Tale of Deadly Obsession in the Amazon.*[2]

Heckenberger has now found cities on the banks of the Amazon similar to those Orellana and Carvajal reported, including 20 outlying towns and villages, spread out over an area of around 7700 square miles. The village of Kuhikugu was likely inhabited from around 1500 years ago to as recently as 400 years ago and had a population of some 50,000 people. Pottery in the area has been carbon-dated to 5500 BCE, 2000 years before Andean or Mesoamerican pottery.

More than 150 such cities have been discovered recently in the Brazilian states of Acre, Amazonas, and Rondônia and in the Bolivian departments of Pando and El Beni, but those are likely "less than 10 percent" of the total, according to archaeologist Martti Pärssinen of the University of Helsinki. This archaeological frontier is living on borrowed time, however. The entire region is scheduled for inundation by a massive hydroelectric project.

At Kuhikugu, large moats, or possibly chinampas with palisades, were built around the communities. One moat that Michael Heckenberger showed writer David Grann was 12 to 16 feet deep, 30 feet wide, and a mile in diameter. The sides of the trench exposed terra preta soils dated to 1200 CE.

Large circular plazas at Kuhikugu extended to 490 feet across, surrounded by dense areas of settlement and urban gardens. Hamlets were crisscrossed by street grids and linked to each other by wide roads on raised causeways, radiating out from city centers in an elaborate regional plan, complete with wooden bridges that spanned large and powerful rivers. Barge and gondola canals passed from the rivers to the interior, parallel to the larger roads, which were embellished with raised curbs and sidewalks. Dams and ponds suggest that the type of fish-farming still practiced by the modern-day Kuikuro was practiced in ancient times. Fields of manioc, corn, and cassava lie at the outskirts of the more heavily settled areas, along with open parklands and working forests.

The population density was 16 to 32 persons per square mile, about the density of Oregon or Colorado in 1990. If the population of the Xinchu region was representative of all of Amazonia, the Amazon's population would have been 42 to 88 million at the time of Orellana's expedition.

In his book *The Ecology of Power: Culture, Place, and Personhood in the Southern Amazon, AD 1000–2000,* Heckenberger says that one of the reasons it has taken so long to find the gleaming cities seen by Orellana is that the monumental architecture practiced by the peoples of the Amazon were not temples, spires, and mounds, but rather more horizontal — the grand plazas, marketplaces, broad causeways, and thriving ports that signified a people less concerned with hierarchy or Sun worship and more concerned with having convivial places to live and work. If the greatest threat to their security was themselves, then the best precaution was to engender a happy and well-cared-for population.

And when it was gone, they left not pyramids in a barren desert but trees and vines, in fertile dark soil.

BOOK II

AGRICULTURE AND CLIMATE

Archaeologists studying the rise of farming have reconstructed a crucial stage at which we made the worst mistake in human history. Forced to choose between limiting population or trying to increase food production, we chose the latter and ended up with starvation, warfare, and tyranny.

— Jared Diamond

The story of civilization that we have been telling ourselves for the past 2000 years is fundamentally flawed. Perhaps because it is missing the voice of so many of the indigenous peoples, particularly those in the Americas, and is so deeply rooted in military empires, our civilization story has glossed over some important lessons we should have learned by now.

The process by which early farming civilizations systematized food production, employing work animals, slaves, irrigation, and the plow, made them more vulnerable to the normal changes in climate that occur periodically — changes due largely to factors beyond human control. History is littered with the bones of civilizations that turned rich soils into deserts, cut down tall forests, and dried up or fouled their precious sources of fresh water.

To prepare ourselves for the changes now coming quickly our way, we should not be following the examples of those who persisted in folly until they exhausted their wealth but rather the examples of those who recognized in fertile soil and biodiversity the best means by which to weather hard times.

10

Making Sand

TWENTY THOUSAND YEARS AGO, as the ice retreated and the climate warmed, the area of fertile soil and suitable growing seasons expanded. Wild game multiplied, drawing human hunters into new regions. About 15,000 years ago, humans began taming dogs, sheep, goats, cows, and pigs. About 8000 years ago, we began saving seeds of emmer wheat, einkorn, barley, flax, chickpea, pea, lentil, and bitter vetch.[1]

In the Fertile Crescent of the upper Tigris and Euphrates rivers, ancient coins bear images of a plow drawn by oxen. Images of planters and plows appear on pottery from Egypt and Anatolia and on rice paper from Japan and China, some more than 14,000 years old.[2] Humans had begun to alter landscapes in very profound ways, clearing forests for fields, expanding villages into cities, and redirecting rivers.[3]

At Iraq's Tell al-Ubaid are found the remains of one of the longer-lived civilizations in history, about 15 centuries. A 2-meter-high mound, 300 by 500 meters, tells us how Ubaidians kept their homes dry from the floods that seasonally refreshed the alluvial plain. Many things were invented here: multi-roomed, rectangular, mud-brick houses, the first temples for the general public, a hub-and-satellite regional urban plan, cuneiform, the wheel, the zodiac, postal delivery, arithmetic, metal sickles, and irrigation agriculture.

The Ubaid glory ended abruptly at 3800 BCE, when the rainy season ceased, the lakes lowered, and shifting dunes arrived for a prolonged

(600-year) visit. The people reverted to semi-desert nomadism and abandoned the area for a thousand years, their "Dark Millennium."

Then, around 3200 to 2900 BCE, fortunes changed again. This is the era known as the Piora Oscillation, an abrupt cold and wet period. Whether it was caused by long-term fluctuations in the cycles of the Earth and solar system or a catastrophic event such as volcanic eruption or asteroid impact is not known.

The Piora Oscillation comes to the present as the floods of Noah and Gilgamesh. The surface of the Dead Sea rose nearly 300 feet (100 meters) and then receded. In the mountains, glaciers advanced, and the tree line descended. Across the Atlantic, the forests of New England suffered a massive loss of hemlock and elm.

The fortuitous return of the rains to Mesopotamia brought the rise of Sumer, a federation of independent city-states. After a breakdown and collapse between 2200 and 2100 BCE, Sumer rose again as the third dynasty of Ur, ushering in a new golden age that lasted another hundred years. Food security gave the Sumerians unprecedented freedom, and they used that time productively to organize public works systems, educate their children, and engage in scientific inquiry that produced remarkable formulae of advanced mathematics, such as the cube root, the Pythagorean theorem (1200 years before Pythagoras), algebra, geometry, meteorology, astronomy, slide rules, multiplication tables, and both sexagesimal and decimal number systems.

So advanced were the scientists of Sumer that recent conspiracy theorists have linked them to the Rhodes-Milner and Bilderberg Groups, Skull and Bones, the Trilateral Commission, the Bavarian Illuminati, the Knights Templar, the New American Century, and the Federal Reserve, all of which apparently grew from special knowledge and power bequeathed by aliens who visited Sumer as tourists posing as gods,[4] something still all too common today.

Sumerian agriculture depended not on spaceships or tractor beams, but on a system of bringing water from the river to the fields by shadufs, canals, channels, dykes, weirs, and reservoirs. The violent floods of the Tigris and Euphrates Rivers meant that canals and reservoirs would build up silt, so the Sumerians routinely dredged the silt from their waterworks and spread it on their fields. After the spring equinox festival season, Sumerian farmers would flood their fields and then drain them. They brought in oxen or cattle to stomp the ground, kill weeds, and deposit fresh manure. Then they raked their fields to remove stones, plowed with oxen, harrowed, and raked the ground smooth again, before planting seed. Whew!

While the flooding, stomping, and raking technique was healthy for many soil microorganisms, the irrigation and evaporation of sun-sterilized topsoil brought about a gradual increase in salinity. By the Ur III period in the 21st to 20th centuries BCE, a more salt-tolerant barley had replaced emmer wheat and lentils as a principal crop.

Salinity, together with some unhelpful foreign interventions, marked the end of the Sumerians, who had made some amazing advances in soil management but neglected the salt problem until it was too late. Or perhaps they learned to brew beer from barley, made salty snacks, and spent too many days glued to sports competitions while their civilization imploded.

After the fall of Ur, there was a decline of social cohesion until Hammurabi forged the Babylonian Empire in 1792 BCE. Then, with Hammurabi's death, the region fell to marauding Kassites, Hittites, and Elamites, only to reunite six centuries later with the re-emergence of the Sumerian city of Assur, beginning the 600-year Assyrian empire.

While there is no evidence that the Romans learned from the unfortunate experience of the Sumerians and salted the fields of their conquered enemies (as has been variously reported), there is some evidence that the Assyrians sometimes did. It was not unlike Iraq igniting the oil fields of Kuwait.

Assyria collapsed in 614 BCE, and by the 11th and 12th centuries CE, the population of Mesopotamia had dropped to its lowest point in

Fig. 3:
Kassite seal,
c. 1830 BCE.

5000 years. Eventually a new Babylonian empire was established, but it succumbed first to Persian conquest and then to the Macedonians under Alexander the Great.

Throughout this tumultuous history, some things in the Fertile Crescent remained relatively constant. The region was dependent on a fragile supply of water; it continuously gave up soil to overgrazing domestic animals, erosion from plowing, and salination by irrigation; and the constant wars over territory required the local farmers and herders to work overtime to feed the large armies that kept arriving and demanding to be fed, and usually not asking very nicely.

And that's how the Fertile Crescent lost its fertility.

In his classic text, *Collapse: Why Civilizations Choose to Succeed or Fail,* Jared Diamond writes:

> Today, the expressions "Fertile Crescent" and "world leader in food production" are absurd. Large areas of the former Fertile Crescent are now desert, semi-desert, steppe, or heavily eroded or salinized terrain unsuited for agriculture. . . . In ancient times, however, much of the Fertile Crescent and eastern Mediterranean region, including Greece, was covered with forest. . . . Its woodlands were cleared for agriculture, or cut to obtain construction timber, or burned as firewood or for manufacturing plaster. Because of low rainfall and hence low primary productivity . . . regrowth of vegetation could not keep pace with this destruction, especially in the presence of overgrazing by abundant goats. With the tree and grass cover removed, erosion proceeded and valleys silted up, while irrigation agriculture in the low-rainfall environment led to salt accumulation. These processes, which began in the Neolithic era, continued into modern times.... Thus, Fertile Crescent and eastern Mediterranean societies had the misfortune to arise in an ecologically fragile environment. They committed ecological suicide by destroying their own resource base.[5]

The Fertile Crescent is not an isolated example. For thousands of years, the inhabitants of the border region between Inner Mongolia and China known as the Mu Us ecosystem had practiced either sedentary agriculture in the sub-humid lowlands or nomadic pastoralism on the high arid plains and forest meadows, each micro-society complementing the others by trade.

With the advent of military dynasties, these societies merged into something more homogeneous, irrevocably altering the natural vegetation by population increases, dry-farming activities, forest clearing, and the impact of wars. The grassland ecology changed to desert so dramatically that sand dunes erased the Xia Imperial Dynasty entirely in 413 CE.

With reduced human impact after the Xia's spectacular fall, some of the forests, meadows, and animal life returned to the Mu Us, only to be cleared again by the Tang and Song dynasties, which burned the forests and planted grain to support their armies. By the 16th century, the dunes were so high that they buried parts of the Great Wall. The Ming and Qing dynasties, exploiting what they called "Mongolian uncultivated land," converted the remaining transitional zones to farmland, channeling the dunes southeast and creating the most promising dune buggy area in Asia today.

Now the remnants of rich farmland, meadows, and forest have become the sands of the Mu Us Desert. They are picked up by intense windstorms and transported across the Pacific, where they rain down on the Alaskan permafrost, darkening its surface and hastening its thaw.

Worse, Mongolia's livestock grew by one-third in the last decade of the 20th century, to reach nearly 33.5 million head. Why so much meat? Because in the preceding 50 years, the human population of Mongolia had tripled.

Just as military ambitions launched the desert of Northern China, so farmers in North America, encouraged by subsidies to grow grain for World War I, plowed up marginal land and grasslands, doubling the cropping area of the southern plains between 1910 and 1920. When the drought years came in the 1930s, the dust turned the Capitol on the Potomac dark at midday and made environmental refugees of those farmers, begging for bread on the Capitol Mall.

From 1968 to 1973, a similar sequence of events crept across the Sahel, the huge transition zone between Sahara and savannas that stretches east-west across Africa, affecting 200 million people. It repeated in Mexico's "Cradle of Maize," the Tehuacán Valley, first in the 1940s and then again in the past decade. Look around. It can be seen unfolding now on the Iberian Peninsula, Andean plateau, and in the American Southwest.

Rather than responding to these calamities with cautious soil preservation and shelterbelt plantings, many governments are either doing nothing or actively promoting overpopulation and overexploitation of sensitive areas. Yet, the lessons of history are clear. This will not last.

From the Library of Congress.

Fig. 4: *Plowing "fencepost to fencepost" in the 1920s.*

11

The Moldboard

THERE WAS SOMETHING SERIOUSLY WRONG with the soil use pattern that arrived with the invention of the earliest plows, cutting the fields of Sumer. We didn't notice it then, and by and large we have not noticed it yet. Even more than the human animal's newfound relationship with cultivated plants and domesticated livestock, the plow dramatically altered the balance between land and sky.

When agriculture was first developed, simple hand-held digging sticks or hoes helped farmers avoid creaky backs. You can see these implements preserved in museums and depicted on ancient coins and pottery. The hoes made furrows. Seeds were sown. Weeds were dug out and covered. In less fertile or sun-drenched areas, the soil was turned to guide richer organic material closer to the surface.

By 4000 BCE, oxen were being used in Mesopotamia and in the Indus Valley to pick up the pace of the farmers. The very earliest plow was the simple scratch-plow, or "ard" — a frame holding a vertical wooden stick that could be dragged through topsoil, using the muscles of an ox. Heavy metal plows, in the form of either solid iron or iron laid over wood, became available by the sixth century BCE in China, leading to a significant population increase there over the following millennium. Greeks, Teotihuacáns, and others introduced the crooked plow, which angled a cutting surface forward. The cutting surface was sometimes faced with bronze, tin, or iron.

Introduced by the Celts in Britain around 4000 BCE, the moldboard

Fig. 5: *Where animal labor was not available, human labor pulled plows, as in this Canadian Doukhobor colony.*

plow added a curved cutting blade called a coulter, knife, or skeith. Coupled with the motive power of oxen, human-slave drag teams, or horses, moldboards could be rigged to cut a number of furrows at one time. With a drag-frame ("landside") or wheels positioned to support the weight, and a skilled plowman, the moldboard could rip along through hard ground at a fast pace.

The moldboard greatly reduced the amount of time needed to prepare a field, and, as a consequence, a farmer with draft animals could work a large area of land. An Amish six-horse team, managed by one 16-year-old Amishman, can plow 10 acres in a day.

An advance on the basic design was the plowshare, a replaceable horizontal cutting surface mounted on the tip of the moldboard. The plowshare spread the cut horizontally below the surface, so when the moldboard lifted it, a wider area of soil was turned over.

Dutch sailors brought Chinese iron moldboard plows to Holland in the 17th century. The Dutch were then hired by the English to drain the East Anglian fens and Somerset moors, and they took with them their Chinese plows. The heavy plows spread to Scotland, North America, and France, where gradual improvements were introduced, until in 1837, John Deere patented the first steel moldboard, "the plow that cut the plains."

One morning in 1846, when Charles Darwin came above deck to observe the passing coast of South America, he ran his finger along the

Beagle's railings and felt the grit of the snowy white dust lying there. It was phosphate, and although he could not know it, it had been picked up from the edge of the Sahara desert some days earlier, traveled on the wind for 5000 miles, and then fallen where he could run his finger through it.

Just beyond the *Beagle*'s railings, the dust fell on the rainforest of the Amazon. It had been doing so for thousands of years. The storms of Brazil annually rain 12 million tons of soil rich in diatomaceous earth, phosphoric acid, nitrogen, and mineral salts formed at the bottom of what had once been Lake Chad.

Twenty miles above, a thin dust of living spores was caught by a gust of solar wind and carried off into the cold of space. Where *that* dust may alight, we know not.

Changing the Paradigm

D RYLANDS COVER MORE THAN ONE-THIRD OF THE EARTH'S LAND SURFACE,[1] and about 70 percent of drylands suffer from degrada-tion, according to United Nations figures.[2] Of irrigated croplands, 30 percent are now coping with degraded soils; of rain-fed croplands, 47 percent; of rangelands, 73 percent. Between one and three billion people work these fields and try to make them produce food, using practices that are almost uniformly soil-destroying, but in many cases encouraged by governments and international lending agencies. As the dryland pop-ulations grow, poverty and disease are increasing. This brings even more unsustainable land practices, and so the downward spiral accelerates.

Every so often, the climate cycles shift again and give us a break. Recent satellite data tell us that the southern Sahara stopped expand-ing after 1980 and seems to be getting moister and greener now. But by elevating population, extending overgrazing practices, or going in with tractors and plows, the nations of that region — there are 11 covered or bordered by the Sahara's sands — could destroy the opportunity they are being given.

Or they could instead plant shelterbelts, and hold the rainfall on contours. They could selectively graze animals to enhance soil fertility while reseeding fields and forests.

We could help nature recover the way nature *likes* to recover.

Our saving grace as humans is that we learn. We have learned, for instance, that desert conditions are not homogeneous, linear, or irreversible. We have learned that just as good agricultural climates can become desert, the reverse is also true. Regional climate can be modified to increase rainfall by reforestation and construction of water-retaining landforms. Drylands ecosystems can be encouraged to sequester carbon, clean water, foster animal diversity, and perform other needed services.

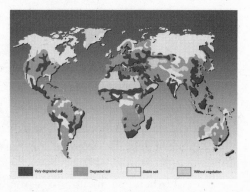

Fig. 6: Degraded soils now extend over more of the landmasses of the Earth than stable soils. After Philippe Rekacewicz, UNEP/GRID-Arendal.

Alternative types of plows have been developed to replace the moldboard, and some recent field trials are promising. One such innovation is the Yeomans keyline plow, which cuts a line through the soil and raises up the top few inches without turning the soil over or exposing it to parching. Australian farmer Darren Doherty says, "It allows the soil to take a deep breath but not leave its mouth gaping open."

The keylining process was developed by Australian stockman P. A. Yeomans in the 1950s. By studying the lay of his land, he noticed that in the usual flow of things, gravity takes water downhill by the shortest route, carrying topsoil and soluble minerals from the ridges and concentrating rich deposits in the valleys. For a farmer or a rancher, what is really needed is the opposite — to distribute soil moisture from the wetter valleys or field indentations out towards the drier ridgelines, and to cover the largest possible area with rich deposits after it rains. Yeomans and his sons experimented with designs for farm implements and techniques of mapping and contour harrowing that would direct rainwater from valley to ridge.

Yeoman's keyline technique follows a twofold pattern — spatial and temporal. The spatial pattern is to follow the topography without following contour, because contour plowing just concentrates water and nutrients back into the valleys instead of distributing them. The better way to deal with runoff in one area is by plowing cultivation furrows to create an opposing drift, away from the runoff direction. Yeomans ranked the primary and secondary ridges and valleys and then used surveying instruments (his son now uses GPS) to mark the line for his tractor to follow, in a herringbone pattern.

The results of keyline cultivation, even on degraded dryland soils, are dramatic. Topsoil is gently raised and loosened with negligible disturbance. Travelling downhill by gravity, water moves from higher valleys to lower-elevation ridges. Rain and air enter the soil and release minerals that chelate and loosely attach themselves to clay particles and humic acid. The released minerals are not water-soluble and are readily available to roots.

Much less disruption of the fungal, protozoa, and microarthropod portions of the food web occurs, meaning less disruption of soil structure, nutrient retention, and potential nutrient cycling. Studies show that a single 6-inch-deep pass with a keyline plow can provide a 12-percent increase in agricultural productivity and a 1-percent increase in organic matter.

Keylining's temporal pattern dictates that plowing occur at the optimal time for carbon to be retained. For grass and pasture growth, that would be when the plants are at their maximum extent, both above- and belowground. Typically, this will be from mid-summer to late fall. Keylining at that time causes shearing and retention of the root systems as soil carbon. In some cases, such as in a tall-grass pasture, it works best if the grasses are grazed to the third-leaf stage, so that there is less folding in of the grass tops and more carbon exudates are retained in the root zone.

The Yeomans Keyline Plow, a $55,000 piece of equipment made only in Australia, resembles the secret "winged" keel shape that helped *Australia II* dethrone the US in the America's Cup sailing regatta of 1983. Operating like a hydrofoil, it has "terrodynamic" horizontal fins at the subsoil bottom of a vertical shank. Three to five of these finned plowshares are mounted on a heavy steel frame and dragged along behind the tractor. A coulter disc precedes the plow shanks, slicing open the upper soil layer to minimize surface disturbance and reduce the energy required to pull the device. The angle of action at the plowshares' leading edge is very slight — only 8 percent compared to typically 25 percent in chisel plows and subsoilers.

Keyline design combines cultivation, irrigation, and stock management techniques to greatly speed up the natural process of soil formation, and results of 400 to 600 tons of topsoil per acre each year are not uncommon. Keylining can deepen topsoil four to six inches annually, and make it three feet deep with repeated annual applications. Keylining also remedies one problem of stock management — compaction — and goes a long way towards eliminating the problem of irrigation, because it captures more of any rainfall and distributes it to where it is most needed.

Fig. 7: *The Yeomans Keyline Plow uses cutting coulter disks to slice a thin groove in the soil, making it easier to pull the shanks and their "terrodynamic" plowshares.*

The millennial effort to design a better plow has not abated. Newer plow designs combine the advantages of no-till with the sustainability of organic production. This marriage does away with no-till's usual downside — the need for heavy chemical applications to control weeds and refresh soluble salts.

The no-till crimper-roller developed by Rodale Institute in Pennsylvania allows farmers to terminate their green cover crops without the use of either tillage or chemicals, and to seed the next crop in the same pass in order to minimize compaction and burn less fuel. To get the same effects, the Keyline Plow frame can be outfitted with seeders and biochar or compost tea applicators, and trailed by a roller harrow.

The no-till crimper-roller is a modified roller with a 16-inch diameter drum and 4-inch blades. In some no-till crimpers, blades are welded onto the drum in a chevron pattern to avoid bouncing and make steering easier. They are angled back from the direction of motion to reduce the amount of soil disturbance.

The blades aren't meant to cut the crop; they push it over and crimp the stems, forcing them into contact with the soil. If a crop is crimp-rolled at flowering, up to 90% of the crop becomes a cover mulch and nitrogen supply. The mulch layer suppresses weeds, or at least delays them, while reducing water losses from the soil surface.

The reason no-till works — or doesn't — is all about biology. If the soil organisms are present and functioning, no-till will work. But

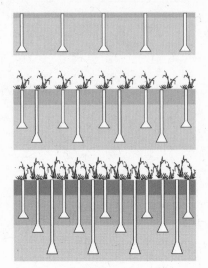

Fig. 8: *Keylining of degraded land progressively for the recommended three years — setting the shanks deeper each year — breaks compaction, increases water retention, and darkens the soil profile with stored carbon and active microbiota.*

just because you stop tilling and applying pesticides does not mean organisms will come back.

Many organic grain producers are returning to the tradition of letting farm animals graze fields that have been harvested, rather than plowing in the stover (leftover dried stalks and leaves). Sometimes farmers will seed a field as it is harvested and let the new plants come up before letting animals in. If kept to appropriate grazing densities, or even better, rotated frequently in high densities, animals remove the excess woody waste that is slow to decompose. Their appetites control competing weeds, avoid the erosion and compaction that a tractor and harrow would invite, and leave enough stubble and ground cover for snow- or rain-trapping over the fallow period. Animals also endow the soil with enteric wastes, as we will discuss in greater detail soon. This assists the recuperation of beneficial soil organisms.

No-till organic management is a paradigm shift of the type we can expect to see much more of as we come to recognize the suicidal folly of the agriculture we've been using since the time of Sumer — and as we come to recognize the happy prospect of an alternative that we had only the briefest glimpse of before it was lost to the world 500 years ago.

If we want to decouple agriculture from desertification, the shapes and sizes of our built environments and the placement of our food production and forests will, in our not-too-distant future, become much less about viewscapes, property lines, or ease of access to the nearest superhighway. Our decisions will instead be about reversing climate change and feeding a hungry planet with the most energy, soil, and water saved or gained along the way.

13

The Amazon and the Ice Age

ABOUT 7000 YEARS AGO, at a point in the solar cycle when Earth should have stopped warming and begun drifting back toward another Ice Age, settlers in China's Yellow River Valley began diverting river water to irrigate wet-adapted strains of wild rice. Over the following 2000 years, the practice spread into Southeast Asia and other parts of China. In these artificial wetlands, vegetation grew, was inundated, died, and decomposed anaerobically (underwater) to produce methane (CH_4), which bubbled to the surface and wafted skyward.

During that same period, as farmers cleared forests to create fields and rice paddies suitable for grazing and working animals, they released both carbon dioxide (from burning) and methane (from rotting plants and domesticated ruminant animals).

As human populations expanded, thanks to the more stable food supply, human wastes also contributed methane to the atmosphere. Methane is 20 times more potent as a heat-trapping greenhouse gas, molecule for molecule, than is carbon dioxide. However, since the population expanded slowly, doubling only every 1000 to 1500 years, wastes were less of a factor for climate moderation than land use changes, which were widespread and substantial, as humans turned grasslands and forests to cultivation. From these alterations, methane concentrations rose approximately 250 parts per billion in the period from 5000 to 500 years ago, and contributed about 70 percent of the interglacial warming experienced in that period. Another major fraction was contributed by anthropogenic CO_2, which added about 40 million tons of heat-trapping carbon to the atmosphere each year, on average, for 7500 years.[1]

In 1089 CE, William the Conqueror ordered a census of people, forest, and farmland in England. From that survey, *The Domesday Book*, we can see that by 700 years before the Industrial Revolution, humans had altered their landscape by removing most of the trees in England and planting crops. Each of the 1.5 million people in England in 1089 required 22 acres (9 hectares) to support the Iron-Age lifestyle.[2] Similar clearances have been demonstrated in Europe and China from much earlier times.

All of that clearing, most profoundly in the fertile lowlands and riverside and coastal areas, increased erosion, compaction, and salination of soils. The world's most fertile soils were the first to be exploited. They were "burned up," rendered incapable of saving carbon during the annual cycles.

In Western Siberia, the world's largest peat bog, roughly the size of France and Germany combined, is thawing for the first time in 11,000 years. That is nice if you are looking for mastodon DNA, but not so good if you were hoping to avoid a climate tipping point. The world's peatlands contain 550 billion tons of carbon (GtC). At warmer temperatures, peat bogs exhale that carbon back to the atmosphere.

Peat bogs can also catch fire, and once they do, the fire is nearly impossible to control until it burns itself out. In 1997, it was estimated that peat and forest fires in Indonesia released the CO_2 equivalent of a third of the fossil fuels burned worldwide in that year (about 2.57 GtC). Those smoldering wildfires are still burning, still igniting forests, and will likely continue until southeast Asia runs out of peat to fuel them, sometime around 2040.

The potential for peat bogs to sequester carbon is large, and rewatering those that have dried out makes fundamentally good sense. But what is actually happening is the opposite. Despite a replacement time scale of a thousand years, peat bogs — about 60 percent of the world's wetlands — are being drained to harvest fuel for electricity and home heating, releasing more greenhouse gases per unit of heat than coal or natural gas.

The presence of methane in bubbles of fossil air found in ice cores was the first scientific proof that humans were directly altering the glacial cycle, but it was largely missed until a decade ago, when William F. Ruddiman, working at the Department of Environmental Sciences at the University of Virginia, noticed a 250-part-per-billion methane anomaly that could not be explained by solar cycles or other natural causes.

Over the past two centuries, methane increases have been far more dramatic than they were in any earlier era of forest clearing and peat burning. World population is now doubling every 60 years, and atmospheric carbon concentrations are rising on an exponential curve. Worse, if all emissions ceased today, it would be half a century or more before the full effect would be felt, and it would take 6000 years to absorb all that man-made carbon back into the oceans.

William Ruddiman picked up another signal in the methane chart of the ice cores, a change that lasted only a few centuries. It was the Little Ice Age, between 1500 and 1750 CE.

What Ruddiman grasped is that the climate response to the hand of man is far more sensitive than had previously been imagined. Every plague and pestilence in history allowed forests to re-emerge — by bringing populations low, fallowing farms, and decreasing burning of peat, coal, and wood. The longer and deeper the plague, the more time the climate had to recover.

Looking at the past 2000 years and factoring in volcanism, Ruddiman could see the fingerprint of plague on changes to atmospheric carbon — the cool Roman Era (200 BCE to 600 CE), the Medieval Maximum (900–1200), and the conquest of the Americas (1500–1900).

After the annihilation of the peoples of the Americas, so great was the burst of vegetation over open fields and elevated ramparts that the carbon drawn from the air to feed this greening upset atmospheric chemistry in an almost mirror image of how the felling of forests — to build those cities and clad them in bright-colored lime-washes and murals — had altered it before.

And that's a good thing.

The significance of the change was not that Louis XIV had to put parquet floors in the Palace of Versailles to keep his feet warm. Rather, the point, to the climate observers now parsing our frightening future, is that humans can alter the atmosphere to take us back to pre-industrial carbon levels — without risky, short-lived, and costly geoengineering gambits such as space mirrors, sulfur aerosols, and fish-suffocating plankton blooms. All we have to do is plant trees, build terra preta soils, and organically store carbon in our planet's terrasphere instead of in its atmosphere.

In 2008, two Stanford University researchers, Richard Nevle and Dennis Bird, bolstered Ruddiman's thesis by examining data from charcoal records from 15 sediment cores extracted from lakes and soil samples from 17 population centers and 18 surrounding sites in Central and South America dating back as far as 5000 years. They looked at ice cores and tropical sponge records for the carbon isotope composition of the atmosphere.

"And it jumped out at us right away," Nevle said. "We saw a conspicuous increase in the isotope ratio of heavy carbon to light carbon. That gave us a sense that maybe we were looking at the right thing, because that is exactly what you would expect from reforestation."

They concluded that reforestation of the Americas pulled up to 50 percent of the total carbon draw-down needed to trigger the Little Ice Age. "There are other causes at play," Nevle said. "But reforestation is certainly a first-order contributor."

14

Predicting Climate's Meander

IN MY 1990 BOOK, *CLIMATE IN CRISIS,* I went over thousands of climate studies that had been reported to that point and included a chapter on the scariest scenarios, as a caution. I called that chapter "Runaway."

We are now a quarter-century later, and that chapter has become the most relevant take-away from my book, indeed, the most salient predictor of our future. I wish it were not true. The overheating scenarios I described, with their positive feedback forcings and tipping points, are an existential crisis we now must face. Denial is a luxury for fools and the soon-to-be-extinct.

What may lie immediately ahead is a drumbeat of human suffering on a scale far greater than any of the natural disasters witnessed in recent years. Sadly, if we still have choices for a rosier outcome, we are watching them slip away with each passing year.

Here is a brief recap of the signs and portents.

In 1824, while working in a Paris laboratory on observations of the Earth, Joseph Fourier described the greenhouse effect for the first time: "The temperature [of the Earth] can be augmented by the interposition of the atmosphere, because heat in the state of light finds less resistance in penetrating the air, than in re-passing into the air when converted into non-luminous heat."

It was a remarkably prescient discovery, given the science of the time. We know now that "heat in the state of light" arrives as high-energy shortwave radiation, able to penetrate atmospheric clouds (or glass windows), and is transformed by contact into infrared, or what Fourier called *chaleur obscure* (non-luminous heat), which attempts to depart as low-energy long-wave radiation, only to bounce back if obstructed (such as by clouds of greenhouse gases). Fourier appreciated the infrared effect from the work of a contemporary, William Herschel, and was quick to realize that how you warm the Earth is the same as how you warm a greenhouse.

Thirty-seven years later, the Irish physicist John Tyndall demonstrated that water vapor is one of the important components of Earth's greenhouse shield. "This aqueous vapour is a blanket more necessary to the vegetable life of England than clothing is to man," Tyndall remarked.

Thirty-seven years after that, Swedish chemist Svante Arrhenius warned that industrial-age coal burning would magnify the natural greenhouse effect. He even provided a number — five degrees Celsius — corresponding with a doubling of atmospheric carbon dioxide.

Thirty-seven years more would pass before carbon emissions from fossil-fuel burning reached one billion metric tons per year, and human population crossed the two billion mark.

Our atmosphere extends a seemingly long ways from the surface of the planet, but the amount of air there is not all that much. If you were to cool that air to a liquid, it would be about 39.2 feet (11.9 meters) deep.

In the 1940s, that air contained about 280 parts per million CO_2 gas by volume (ppmv), which was a 10-ppmv jump over the norm of 270 that had prevailed since the Little Ice Age, and a few dozen more ppmv beyond the pre-agricultural norm. If you were to cool that 1940 CO_2 volume to a liquid, it would be about the thickness of a page of a newspaper, all over the planet.[1]

By the 1950s, measuring equipment had improved to the point where Gilbert Plass could detail the infrared absorption of various gases; Roger Revelle and Hans Suess could show that seawater was incapable of absorbing the rate of man-made CO_2 entering the atmosphere; and Charles David Keeling could produce annual records of rising atmospheric carbon levels from observatory instruments in Hawaii and Antarctica.

Two years after Keeling's first climb up the slopes of the Mauna Loa volcano, human population crossed the three billion mark.

In 1965, the US President's Scientific Advisory Committee warned Lyndon B. Johnson that the greenhouse effect was a matter of "real concern." In 1975, climatologist Wallace Broecker coined the term "global warming" and began warning that sudden climate shifts were not historically unprecedented.

The decade that followed brought a spate of in-depth inquiries by scientific bodies, government committees, and the United Nations. In 1988, the Intergovernmental Panel on Climate Change (IPCC) was formed to report the collected findings in a nonpartisan way.

By 1987, 20 years after Keeling's fateful ascent of the volcano, and nearly a century after Arrhenius's calculations, human population had reached five billion.

Following the release of the first IPCC report, the governments of the world convened the 1992 Earth Summit in Rio de Janeiro. There they agreed to the United Nations Framework Convention on Climate Change, with a key objective of "stabilization of greenhouse gas concentrations in the atmosphere at a level that would prevent dangerous anthropogenic interference with the climate system."

Three years later, the second IPCC report described how "dangerous anthropogenic interference" with the climate system was already occurring. Meanwhile, virtually no progress was made on any of the goals that had been set for emissions reductions. Business-as-usual had taken control, and the denial machine had gone into overdrive.

In 1999, human population reached six billion.

In 2001, the IPCC Third Assessment Report issued even stronger warnings, based on new evidence of dangerous anthropogenic interference. In response, the UN enacted the Kyoto Protocol to mandate reductions, with the target of returning the world to its 1990 emissions level by 2012. These voluntary reductions failed to significantly alter the growth of fossil fuel consumption, or slow the admission of powerful new members to the greenhouse-gas-producer family, such as China, India, Indonesia, and Brazil. World population was still growing, and everyone seemed to want a house filled with the latest consumer appliances, plus a two-car garage, a chicken in every pot, and a pot roast in the oven.

By 2006, atmospheric CO_2 had risen to 380 ppmv, the thickness of a few sheets of newspaper if condensed and laid across the ground.

In 2007, carbon emissions from fossil-fuel burning reached 8 billion metric tons per year (8 GtC/yr), with a comparable amount being generated by animals raised for slaughter. The IPCC's Fourth Assessment Report concluded, almost anticlimactically, that it was now certain that humanity's emissions of greenhouse gases are responsible for climate change. At framework negotiations in Bali, world governments agreed to the two-year "Bali roadmap" aimed at hammering out a new global treaty by the end of 2009.

If your species loses its economic underpinnings, your position on the economic development chart could fall back to where you were in 1930, or 1830, or even 1330, but you are still around as a species, assuming you don't totally lose your cool and just nuke everything in sight on your way down.

In contrast, if your species loses its climate underpinnings, it's "Game over, man." You not only take down the higher vertebrates, homo included, but everything alive on this third rock from the Sun, potentially even the microbes in deep caves and ocean depths. Earth, meet Venus.

On May 19, 2009, Woods Hole Research Laboratory and the Massachusetts Institute of Technology released a study involving more than 400 supercomputer runs of the best climate data currently available. Conclusion: The effects of climate change are twice as severe as estimated just six years ago, and the probable median of surface warming by 2100 is now 5.2 °C, compared to a finding of 2.4 °C as recently as 2003. Moreover, the study rated the possibility of warming to 7.4 °C by the year 2100 (and still accelerating thereafter) at 90 percent — in spite of our feeble efforts at "cap and trade," "contraction and convergence," or a "clean development mechanism."

Another report, released in 2009 by the Global Humanitarian Forum, found that 300,000 deaths per year are already attributable to climate-change-related weather, food shortages, and disease. That

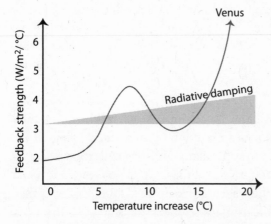

Fig. 9: *Recent studies with global climate models suggest that new equilibrium could be found between 5 and 10 degrees warmer, but that at some point the natural damping effect is overwhelmed and a runaway greenhouse warming transforms the climate of Earth to something more resembling Venus. After a slide by Hans J. Schellnhuber, "Terra Quasi-Incognita: Beyond the 2-Degree Line."*

figure could be called our baseline, or background count — of the 20th-century-long experience of a temperature change of less than 1 °C.

These two reports set the stage for the findings released by the US Advisory Committee on Global Change Research, *Global Climate Change Impacts in the United States.*[2] The authoring team was headed by senior climate scientists at the National Oceanic and Atmospheric Administration and included half a dozen government agencies and laboratories, and senior researchers from a dozen universities. Just reading the introduction is refreshing, because it cuts through so much of the tone-deafness that passes for public debate these days.

While limited solely to the continental United States and Hawaii, the *Impacts* report takes the projections for the coming decades to about as fine a grain as can be seen, given the behavior of interrelated and reciprocating climate systems undergoing rapid destabilization. Here are some findings that really gave me a 1990 *déjà vu* feel:

- The European heat wave of 2003 [with more than 30,000 heat fatalities] is an example of the type of extreme heat event that is likely to become much more common. If greenhouse gas emissions continue to increase, by the 2040s more than half of European summers will be hotter than the summer of 2003, and by the end of this century, a summer as hot as that of 2003 will be considered unusually cool.

- Recent findings indicate that it is very likely that the strength of the North Atlantic Ocean circulation will decrease over the course of this century in response to increasing greenhouse gases. . . . The best estimate is that the strength of this circulation will decrease 25 to 30 percent in this century, leading to a reduction in heat transfer to the North Atlantic. It is considered very unlikely that this circulation would collapse entirely during the next 100 years or so, although it cannot be ruled out.

If we burn all the fossil fuels (including the gases released by fracturing the oil shales, steam-heating the tar sands, and tapping all the deepest ocean deposits), releasing many gigatons of carbon, there's a chance the Earth's temperature will surpass Cretaceous levels of about 65 million years ago, and the seas could reach 38 °C (100 °F), hotter than the human body. Today's sea surface temperature is 16.4 °C, (61.5 °F). Acidity, which reflects the amount of greenhouse gases absorbed by the oceans without being precipitated out, is higher than it has been in 65 million years.

While it would be comforting to sit back and bask in the revealed glory of 20-year-old predictions come true, I am instead left with a deep and abiding sense of foreboding that colors everything I say and do now. Finding ourselves trapped in a burning building, we have to search out and consider any potential escape routes, and quickly. We can ignore the deniers, because they are only impediments to us now. Our survival, and that of the experiment of life on a blue water world, depends on our ability to keep clarity and resolve as all around us the flames, smoke, and panic are rising.

Charles Mann's description of the decimation of the Americas haunts my dreams. "The pall of sorrow that engulfed the hemisphere was immeasurable. Languages, prayers, hopes, habits, and dreams — entire ways of life hissed away like steam."

That is the reason we keep reading through all this doomer porn, toss in our sleep, get bags under our eyes, and curse our fate.

We should not forget that what we need to do in order to extricate ourselves — *garden Earth* — is also going to make our lives vastly better then they otherwise would have become, and our children's lives will be still better, although quite a bit warmer for a while.

BOOK III

CAPTURING CARBON

Progress! You politicians are always talking about it.
As though it were going to last. Indefinitely.
More motors, more babies, more food, more advertising, more
money, more everything, forever. You ought to take a few lessons in
my subject. Physical biology. Progress indeed! What do you propose
to do about phosphorus, for example?

— Lord Edward Tantamount
in Aldous Huxley's *Point Counter Point* (1928)

Huxley's point about phosphorus, which he also made in *Brave New World* (1932) with a Ministry of Crematoria and Phosphorus Reclamation, is that we live on a finite planet and yet we scatter life-giving minerals to the oceans and the winds as if they were infinite. Percentages of nitrogen, phosphorus, and potassium are prominently displayed on every package of fertilizer, and arguably our ability to obtain nitrogen from the air and mine phosphorus and potassium enabled us to reach a population of seven billion. Conventional agriculture likely has enough potassium to last several centuries, but, as Huxley warned, phosphorus is a different story. The carbon and nitrogen cycles, as we shall see, are even more stressed, more out of balance, and more population-threatening.

What Huxley's Lord Tantamount was chiding politicians about is their stalwart devotion to abstractions like economic growth or progress while turning a blind eye to the needs of natural processes that support all human activity. Whether we chart the phosphorus cycle, mountain snows, or fish populations in coral reefs, the trends of the past decades are ominous and alarming. Whatever we are doing, it is not progress.

15

Carbon Farming

A S CONVENTIONAL FARMING SPREAD FROM SUMERIA to the rest of the world, the soil biology deficit steadily grew. Measured in carbon, we can put this deficit in 2010 at 30 to 75 percent, depending on location, pre-existing soil type, climate, terrain, drainage, and land use or abuse.

The good news is that solely by changing our farming practices to ones more like the ancients we can take carbon out of our air while, on average, incorporating a layer of carbon into our soil equal to 0.081 inches (2.1 millimeters) over the 8.5 percent of the area of the Earth that we currently use to make food. While degraded soils are the most receptive carbon sink, nearly all soils can be improved by adding carbon.

The process of restoring the soil's biological life, which we'll call carbon farming, not only draws carbon from the atmosphere to the soil, it also increases biomass productivity, increases food production, improves nutritional value, helps water purification, reduces energy and fertilizer requirements, controls pests and weeds, and significantly increases biodiversity.

Many of the soils of the world — in the American plains, the Australian outback, the Ural foothills, and the Fertile Crescent — once had carbon content of up to 20 percent, before the advent of the plow and goat. They are typically at 0.5 to 5 percent today. Because fertility varies, it is possible that we can get more return for our carbon investment in some places than others. Generally, the more "worn out" the soil, the more carbon it can take back. This is good news for Africa, Australia, and other areas of depleted soils.

Soil scientist Rattan Lal of Ohio State University found that with better carbon management practices, soils in the continental US could soak up 330 million tons of carbon each year, enough to more than offset the emissions from all the cars in the US, while improving food production by 12 percent.[1] Lal says the ultimate potential for soil carbon uptake is one billion tons — a gigaton — per year (1 GtC/yr).[2] This contrasts with the 800 gigatons of carbon in the atmosphere, about half of that being anthropogenic. Lal's estimate may be low, but if it were accurate, and if man-made emissions could be brought down to zero, carbon farming could cut carbon in the atmosphere by one part per million every four years. We are currently raising it by nearly two parts every year, so cutting emissions is still the key to returning to a safe climate.

We have no shortage of carbon to draw upon. The atmosphere carbon pool increased by 3.3 GtC/yr during the 1980s, 3.2 GtC/yr during the 1990s, and an average of 4.1 GtC/yr between 2000 and 2005. The concentration of carbon dioxide has increased 39 percent from the pre-industrial level of 280 ppmv to 390 ppmv today. We can reverse that process using biochar in combination with carbon farming and tree planting. Few other human activities can make that claim.

The uptake of CO_2 by natural sinks (terrestrial biosphere and the ocean) has been about 50 percent of total emissions each year, which is to say that Gaia, the geo-chemico-biological stasis of the planet, has been able to absorb half of what we gave her, but no more. Of that, the uptake by the terrestrial biosphere (soils and trees) was about 1 gigaton of carbon per year higher in the 1990s than during the 1980s, which

Calculating Carbon

One gigaton is a billion tons. An English ton — 2000 pounds — is 10 percent less than a metric ton — 1000 kilograms. For simplicity, I use them interchangeably.

The molecular weight of carbon dioxide (C = 12 + O = 16 + O = 16) is 44. Carbon alone is 12. Therefore the molecular weight of Carbon Dioxide Equivalent (CO_2e) is calculated at 3.67 times the weight of carbon alone.

indicates that Gaia has been stretching a little to accommodate our higher atmospheric emissions. However there are signs that she has reached her limits.

The total amount of organic carbon in all of Earth's soils is estimated to be 3200 GtC, which constitutes about 70 percent of the carbon in terrestrial ecosystems.[3] The carbon dioxide stored in plants, 650 GtC, is 81 times all annual man-made CO_2 emissions (8 GtC). Thus, diverting only a small proportion of this plant growth to cycles that would hold carbon out of the atmosphere — in "farmed carbon," trees, houses, bamboo furniture, wooden ships, or biochar — would begin arresting climate change.

As contrasted with the crystalline, recalcitrant carbon in biochar, the usual form of carbon molecules found in biological systems is called "labile" because it is very active and easily attaches to other organic

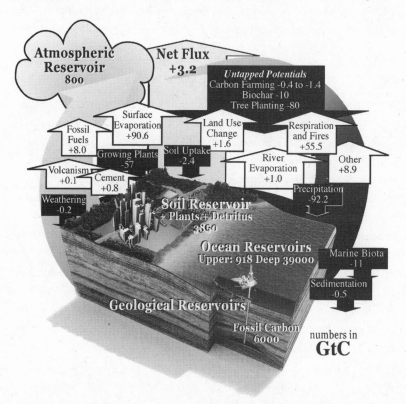

Fig. 10: *The carbon cycle. "GtC" represents a gigaton (one billion tons) of carbon.*

molecular chains in solid, liquid, and gaseous forms. Plants and animals, ourselves included, need labile carbon to form the building blocks of cellular tissue, including the DNA nucleoprotein helices in all living things.

There is a big difference in the amount of carbon sequestration realized from recalcitrant versus labile carbon management practices. Recalcitrant carbon sequesters 25 to 50 percent of the carbon of its source feedstock for 1000 years. By comparison, making labile carbon by biological decomposition (plowing under) or torrifaction (stubble-burning) sequesters less than 10 to 20 percent for 5 to 10 years.

Johannes Lehmann of Cornell University looked at how fast the greenhouse balance could be shifted using biochar. Depending on various assumptions of how it was made and applied, he estimated the global storage capacity at 224 GtC for cropland and 175 GtC for temperate grasslands; he did not examine forests. Said Lehmann, "These total sequestration opportunities are high and approach levels for total

The Carbon Balance

The present atmospheric concentrations of greenhouse gases are 800 gigatons of carbon (GtC), 3000 gigatons of carbon dioxide (GtCO$_2$), or 3300 gigatons of carbon dioxide and equivalent gases (GtCO$_2$e). Each year, human activities add more than 8 GtC, 30 GtCO$_2$, or 33 GtCO$_2$e.

The annual air-earth-air carbon cycle moves 58 GtC or 215 GtCO$_2$. Of cyclical carbon stores, 21 percent is in plants and marine biota (600 GtC), 27 percent is in the atmosphere (800 GtC), and 52 percent is in the soil (1500 GtC).

The oceans hold another 40,000 GtC in dissolved carbonates that normally do not cycle very much — at cold temperatures. But as the oceans warm, the carbon will begin to return to the atmosphere as CO$_2$ and methane (CH$_4$). Presently the oceans absorb 3 GtC/yr, but that is declining.

Atmospheric components are also defined in parts per million by volume. The threshold at which Earth shifts to a 5-degree-warmer equilibrium is considered to be 95 ppmv C, 350 ppmv CO$_2$, or 393 ppmv CO$_2$e. At present the atmosphere contains 100 ppmv C, 390 ppmv CO$_2$, or 436 ppmv CO$_2$e.

It takes 2.12 GtC to add 1 ppmv CO$_2$. At present, atmospheric carbon grows by 4.5 GtC/yr from human activity, resulting in a 2-ppmv increase each year.

C in plants," meaning that the soil uptake potential for biochar could be as high as the total weight of all the biomass on Earth. Lehmann estimates 12 percent of total fossil fuel emissions could be offset annually if slash-and-burn techniques in tropical cultures were replaced by slash-and-char (burying the smoldering biomass during and after the burn), and a comparable amount could be offset by diverting wastes such as forest residues, mill residues, field crop residues, or urban garbage, to make biochar.[4]

Carbon farming on degraded soils involves conversion from plow tillage to organic no-till farming, with crop residue mulch and cover cropping, integrated nutrient management and manuring, and complexing of cropping with rotational grazing and agroforestry.

A farm that switches to organic no-till can sequester 1 to 4 tons of organic matter per acre per year. By employing perennial polycultures, rotated pastures of grazing animals, trees, and wild plant strips, that amount can be doubled or tripled. Plantings of appropriate species,

Types of Soil Carbon

Soil uptake of carbon dioxide falls into two distinct categories: soil organic carbon (SOC) and soil inorganic carbon (SIC). Carbon sequestration begins with photosynthesis, which inhales CO_2 to create biomass. Part of that biomass is processed into SOC by root hairs left in the ground and by microbes that die and leave their remains. The SOC quickly degrades through physical, chemical, and biological processes, and much of it eventually off-gases back to the atmosphere. Increasing the SOC pool size has a positive impact on net carbon in the soil, and on stabilization of soil structure. It increases water and nutrient retention and species diversity within the soil, and it speeds the recycling of nutrients.

The sequestration of SIC occurs when CO_2 becomes carbonic acid and precipitates carbonates of calcium and magnesium. Leaching of bio-carbonates into the groundwater transfers them away from the root zones where otherwise they might be taken up by plants or soil organisms. How far and fast SIC travels depends on a range of soil and environmental factors such as the amount and nature of the clay that is present, groundwater properties, the depth of the soil, and the availability of binding elements (e.g., nitrogen, phosphorous, sulfur, calcium, magnesium).

stand management, and polyculture guilding can be applied to wood-lots and forests to bring them into the carbon farming regime.

To recover one percentage point of soil organic matter, around 12 tons of organic matter per acre would have to enter the soil and remain there. Because about two-thirds of organic matter added to agricultural soils will be decomposed by soil organisms and plants and given back to the atmosphere, in order to add permanently 12 tons, a total of 36 tons of organic matter per acre would be needed. This cannot be done quickly, or it just washes or evaporates away. A slow process is needed.

If the recuperation of soil carbon becomes a central goal of agricultural policies worldwide, it would be possible and reasonable to set as an initial goal the sequestration of one half-ton per acre-year. As soil conditions improve, and erosion and pests decline, and the land comes back into balance, that target goal could be increased. Farming this way globally could sequester about 9 percent of the current total annual human-made emissions of 8 GtC.

The gains in fertility would mean that chemical fertilizers could be (and should be) eliminated where carbon farming is practiced. Reducing emissions of nitrous oxide from fertilizer (equivalent to approximately 8 percent of annual human-made greenhouse gases) and reducing the transportation and energy impacts of fertilizer production could shave off another few percent of global emissions.

Moreover, if organic waste is returned to agricultural soils in the form of compost, then methane and CO_2 emissions from landfills and wastewater (equivalent to 3.6 percent of man-made emissions) could be significantly reduced.

Even a modest start would have the potential to offset global greenhouse gas emissions by approximately 20 percent per year. And we are talking only about the first 10 years. After that, we can progressively increase the reincorporation of organic matter into soils. By mid-21st century, we could increase the total world reservoir of carbon in the soil by two percentage points, and possibly more.

Exponential curves do not exist in nature, for long. If population trends at the beginning of the 21st century were to continue without limits, agricultural output would have to double by mid-century. We know this to be impossible. By 2050, the per-capita available farmland will be one-third of what it was a century earlier. Our agriculture will either have to become inconceivably more productive than it is now or radically shift its priorities away from non-food products and wasteful but profitable industrial practices.

For the sake of our immediate food needs, if for no other reason, we have to change the way we grow calories, and perhaps also the way we consume them. Gaian ecologist James Lovelock thinks we will develop synfoods — the *Soylent Green* scenario. The UN Food and Agriculture Organization thinks we will just apply (apparently unlimited) fossil fuels to ramp up the Green Revolution faster, the way China seems to be doing. Singularitarians like Vernor Vinge and Ray Kurzweil are banking on some kind of trans-humanist scenario, wherein computing power surpasses human brainpower to the point that we just port ourselves over to something less resource-demanding than our old biological substrate — nanobots, for instance, or pure cyberspace, for the more daring.

It would, of course, help to limit our population growth and to reduce the consumption of meat and dairy products. It takes 16 pounds of grain protein to yield 1 pound of beef protein; for pork, the ratio is 6 to 1; for chicken, 3 to 1. We also need to be much more circumspect about how transitional biofuels are produced — assuring that they complement, rather than compete with, our food supply. We need to rethink the role of soy, maize, and other crops being grown for purposes other than food. By shifting priorities, we can maximize our available caloric intake while we await either the singularity, synfoods, or the transition to a sustainable agriculture of the type discussed in this book.

If you are just not into the Freon diet or can't get with a techno-geek synfood program, then returning to healthy foods produced in abundance from healthy land might be more to your liking. Of course, you'll also want farms to bring our atmosphere to a "normal" 270 ppmv of CO_2-equivalent greenhouse gases, or there really is no future, or at least there will be no one around to appreciate it.

Fortunately, it is possible to feed all those new people not by doubling how much we grow but by changing what we grow, and simultaneously changing how we grow it.

Understanding Soil

MYCOLOGIST PAUL STAMETS is fond of telling audiences that all animals, ourselves included, are descended from fungi, but that we took a slightly different road in our evolutionary choices. Animals chose to have internal organs perform most of our biological processes. Fungi chose otherwise.

When we left our fungi family, we kept many of the same traits that our ancestors had. We exhale carbon dioxide and inhale oxygen, just as fungi do. Our fertilized eggs reproduce a coded sequence for cell differentiation that births and matures our offspring outside of our bodies, just as fungi do.

But, after we split from fungi, fungi went underground and we went overground in the way we choose to locate and digest our food. We elected to internalize the eating process with a nose, mouth, esophagus, stomach, and gastrointestinal tract. Fungi chose to do it externally, using thin threadlike cells that form networks of subsurface mycelium that excrete digestive enzymes into the host substrate. They absorb the digested material through their cell walls and excrete the soluble nutrients.

Fungi appeared on land about 1.3 billion years ago. Plants followed 600 million years later. The plants would not have succeeded without entering into a symbiotic relationship with fungi. Animals, insects, and plants that paired with fungi made it through the great extinction events — the bottlenecks of our genetic history. It was like buying protection from the Godfather. Fungi are history's Mafia.

Fungi munch on rocks. They do this by producing acids that dissolve out minerals. They take calcium, phosphorus, potassium, manganese, and many other elements out of the rock mantle of the Earth and hand them off to the roots of plants.

While we have five or six layers of skin to protect us from infection, fungal mycelia have only one. A mycelial strand, the largest organism in the world by weight or volume, is only one cell wall thick, and it is surrounded by millions of hungry microbes that want to consume it. Fungi protect themselves by attending to the needs of the soil community. Fungi are the grand molecular disassemblers in nature. They provide for their grandchildren, the plant and animal community, by supplying nutrients in forms that plants and microbes want and need. They are the food bank.

Much of this is understood, or intuited, by traditional farmers in widely varying cultures. Western science has been slow to catch on. Thread-like structures within the root cells of certain vascular plants were observed as early as 1829, and by 1885 it was widely known that fungal hyphae, the branching filaments of mycelium, are the mechanism by which nutrients are transferred from humus to the rootlets of plants.

In 1943, Lady Evelyn B. Balfour wrote in *The Living Soil*, "Mycorrhizal association plays a critical part in the nutrition of pines, the fungal activity producing substances that stimulate the growth of lateral roots." Visiting a coniferous forest at Wareham, England, she observed that trees that had been attacked by some soil organism were not improved by "the application of nutrient salts of nitrogen, phosphorus, and potash" but were rapidly cured and resumed healthy growth when subjected to "the application of composts made by fermenting vegetable wastes with 1 percent nitrogen in the form of dried blood."

Balfour concluded:

The action of the compost is not due to the plant nutrients it contains, but to its biological reaction which has the effect of fundamentally modifying the soil microflora. The restoration of health and vigour in the pine seedlings, brought about by compost treatment, is not due to the direct action of the compost on the plant roots, but to indirect action due to stimulation of mycorrhizal and other soil fungi, and the removal of the substance previously inhibiting their growth.

One of the foundations of Balfour's understanding was laid when Franklin Hiram King sailed from North America to China at the start of the 20th century. King, a former chief of the soil management division of the US Department of Agriculture, went to find out how peasant farmers in China, Korea, and Japan could work the same fields for thousands of years without destroying fertility and without applying artificial fertilizer. In 1911, King published his classic, *Farmers of Forty Centuries: or Permanent Agriculture in China, Korea and Japan*. It described composting, crop rotation, green manuring, intertillage, irrigation, drought-resistant crops, aquaculture and wetlands farming, and transport of human manure from cities to rural farms.

King wrote:

The human waste must be disposed of. They return it to the soil. We turn it into the sea. Doing so, they save for plant feeding more than a ton of phosphorus (2712 pounds) and more than two tons of potassium (4488 pounds) per day for each million of adult population. The mud collects in their canals and obstructs movement. They must be kept open. The mud is highly charged with organic matter and would add humus to the soil if applied to the fields, at the same time raising their level above the river and canal, giving them better drainage; thus are they turning to use what is otherwise waste, causing the labor which must be expended in disposal to count in a remunerative way.[1]

King's publications inspired later work by Sir Albert Howard, Rudolf Steiner, Mokichi Okada, and J. I. Rodale. At the core of the emerging agricultural philosophy that would be called "organic" (by Rodale), "biodynamic" (by Steiner), and "natural" (by Okada) were the underpinnings that in nature there is no such thing as waste and that everything living depends on other living things, not just chemicals. As Okada put it, growing food "involves developing a more intuitive relationship with the natural world, and it means listening, respecting, and responding to, rather than dictating, the needs of nature."

To many, Howard is regarded as the founder and pioneer of the organic movement. Raised on an English farm, he served as a mycologist in the Imperial Department of Agriculture for the West Indies, taught agriculture in England, and then moved to India, where for 26 years he directed several agricultural research centers. In 1943, back in England again, he published *An Agricultural Testament,* which described his theories of building compost piles, recycling waste materials, and creating soil humus as a "living bridge" between soil life (such as mycorrhizae and bacteria) and healthy crops, livestock, and people.

Howard became the center of the mid-20th-century conflict between the traditional farming methods described by King and the methods of Nobel-prize-winning "organic chemists" Carl Sprengel and Justus von Liebig, who advocated the practice of fertilizing principally with nitrogen, phosphorus, and potassium salts, rather than encouraging a living soil. Howard's position was, in the words of biologist Janine Benyus, that "it is life that best creates the conditions that are conducive to life."

Spurred by Sprengel and Liebig's Nobel work on chemical fertilizers, Fritz Haber went to work to separate nitrogen from the air and in the summer of 1909 succeeded in producing ammonia, a liquid nitrogen compound, drop by drop, at the rate of about a cup every two hours.

The German chemical company BASF then assigned Carl Bosch the task of scaling up Haber's apparatus. Haber and Bosch were later awarded their own Nobel prizes, in 1918 and 1931, and BASF went on to fame and fortune by turning the nitrogen in air into munitions for World War I.

Although free nitrogen comprises 78 percent of the atmosphere, it is inert in its gaseous state, and neither plants nor animals can absorb it. Apart from the small fraction that is fried apart by the occasional volcano or lightning strike, we mostly get the nitrogen we need through the action of a small group of specialized bacteria that can break the triple bond between two nitrogen atoms, known as nitrogen fixation, creating reactive nitrogen. These powerful bacteria, a diverse and gregarious lot, crawl on land and swim in fresh and salt water. In gardens and forests, they strike up relationships with the roots of legumes and supply nitrogen to nematodes.

Before the past century, the amount of reactive nitrogen produced in the world was balanced by the activity of another bacterial group that converts reactive nitrogen in plants and soil back to nitrogen gas in a process called denitrification. In only one human generation, that delicate balance has been transformed completely. The Haber-Bosch process now produces 100 million tons of nitrogen fertilizer per year, mostly in the form of anhydrous ammonia, ammonium nitrate, and urea. Today combined industry produces more than 400 billion pounds of reactive nitrogen per year, an amount at least double that of all natural processes.

The global warming potential of nitrogen (commonly released as nitrogen oxide, N_2O), with its century-long average hang time in the atmosphere, is 298 times that of CO_2.

With the advent of ammonium fertilizers, the Sprengel-Liebig Law of the Minimum — supplying just chemical salts that encourage growth rather than building soil microbiota — became a widely accepted agronomic principle and the basis for the Green Revolution. Meanwhile, for a lost century, Howard's Law of Return — coupling agricultural production to a healthy soil food web — was known only to organic farmers and not even taught in most schools of agriculture.

Since the mid-20th century, many agricultural systems have shifted from labor-intense to more capital- and chemical-intense. Prior to that

Biofuel Nitrogen

The rise in atmospheric nitrogen isn't just coming from fertilizer production, soluble chemical runoff, or manure. As biofuel industries ramp up to jump the gap opened by peak oil and gas, significant volumes of N_2O are being given off by fermentation and transesterification.

"Renewables," rather than improving the greenhouse balance, are contributing to the rise in nitrogen emissions — up to 62 percent for ethanol made from sugar cane; up to 13 percent in North America for diesel produced from rapeseed; 9 percent in the European Union; and 72 percent in Eastern Europe. Soy biodiesel could increase greenhouse impact from between 30 percent (in the North) to 44 percent (in the South) over the fossil fuels it's replacing.[2]

time, most crops were produced largely without the use of chemicals. Insect pests and weeds were controlled by crop rotations, destruction of crop refuse, timing of planting dates to avoid high pest population periods, mechanical weed control, and other time-tested and regionally specific farming practices.

Although these practices are still in use, government policies, changes in technology, and changes in prices have led to the development of today's "conventional" agriculture. Today, food production is characterized by mechanization, monocultures, the use of synthetic chemical fertilizers and pesticides, and an emphasis on maximizing productivity and profitability. We do well to remember this whole system is only a scant 60 years old.

This type of agriculture is unsustainable because it destroys nearly everything plants depend upon. Soil fertility is declining due to erosion or oxidation of organic matter, water supplies are being depleted and polluted, finite fossil energy supplies are being exhausted, and the economies of rural communities are left in shambles as agricultural outputs are shipped to distant markets.[3]

World cropland per capita is presently only 0.6 acre (0.24 hectare). In China, it is only 0.2 acre (0.08 hectare) — only 15 percent of the 1.25 acres (0.5 hectare) considered a bare minimum to sustain the diverse diet of North America and Europe. The shortage of productive cropland, decreasing land productivity, and the enormous waste and imprecise management associated with industrial-scale processes are responsible for most of the recurrent global food shortages and associated human malnutrition.[4]

Globally, agriculture accounts for about 14 percent of the total greenhouse gas emissions,[5] including 47 percent of methane emissions and 84 percent of the nitrous oxides.[6] From 1990 to 2005, global agricultural emissions increased by 14 percent, N_2O from soils by 21 percent, N_2O from manure management by 18 percent, and methane from enteric fermentation by 12 percent.[7]

Sustainable agriculture used to be defined as agriculture that can be pursued indefinitely because it does not degrade or deplete the resources that it needs to continue. That is no longer good enough. Humanity now confronts a critical challenge: to develop methods of agriculture that sequester carbon, enhance soil fertility, preserve ecosystem services, and use less water and hold more of it in the landscape, while productively employing a steadily compounding supply of human labor. These are the prerequisites of sustainable agriculture.

Fortunately, for the past half-century, a group of pioneers have been preparing the agriculture of the future, and their ideas are now moving to center stage. Organic no-till, permaculture, agroforestry, perennial polycultures, aquaponics, biointensive and biodynamic farming, and natural farming — long considered fringe ideas — are now converging as serious components of a sustainable agriculture.[8]

Sir Albert Howard's ideas found a few important advocates, including the American businessman and publisher J. I. Rodale, who purchased a farm near Allentown, Pennsylvania, and began experimenting with organic techniques. In 1942, Rodale began publishing *Organic Gardening* magazine, which grew in popularity steadily through the second half of the 20th century. Today the Rodale Institute still farms the land that J. I. and his children and grandchildren tilled, and Rodale Press still publishes *Organic Gardening,* and a host of magazines and books that share the Rodale/Howard philosophy of abundance through soil-building.

In 1978, the words "permanent agriculture," from the subtitle of both J.R. Smith's 1929 book, *Tree Crops* and F.H. King's 1911 book, were combined by Bill Mollison and David Holmgren to title their seminal book, *Permaculture One.* Mollison expanded the theme in 1988 with the more thoroughgoing *Permaculture: A Designers' Manual,* and Holmgren applied the concepts of "cultivated ecology" to the emerging crises of energy and climate in his 2002 book, *Permaculture: Principles and Pathways Beyond Sustainability.*

Beginning with agriculture and then extending outwards to reach every aspect of human activity and the built environment, Mollison and Holmgren provided a systems approach to designing human ecologies that mimicked the relationships found in natural biomes. They acknowledged permaculture's indebtedness not only to King and Howard, but also to Masanobu Fukuoka's *The One-Straw Revolution: An Introduction to Natural Farming* (1978), which described the perennial food landscapes and no-till organic polycultures the author had spent a lifetime developing in Japan; and to brothers Eugene P. and

Howard T. Odum, authors of *The Fundamentals of Ecology* (1959) and many other foundational works in systems thinking and environmental economics.

The Odums provided the first empirical studies of energy flow-through systems, developing early computer modeling of humans as part of ecosystems. Later, H. T. Odum coined the words *emergy* (the total solar energy used directly and indirectly to make a product or service) and *transformity* (the energy of one type required to produce a unit of energy of another type), and, contemporaneously with Kenneth Boulding and Charles A. S. Hall, introduced the notion of net energy analysis, which Hall further developed into energy return on investment (ERoI).[9]

With all this impressive work by sustainable-agriculture pioneers, some new agricultural approaches are coming into view, but they are not magic elixirs. While optimized farming practices can increase the capacity of the land to produce over the long term, they cannot be considered in isolation; a robust solution to our continued existence on this planet must include adopting sustainable lifestyles and maintaining human population at sustainable numbers. Agriculture alone cannot keep up.

The shift to biologically based farming (as opposed to chemically based) conceives of the whole farm as an organism — a self-contained integration of crops and livestock, nutrient recycling, soil maintenance, and the health and well-being of crops, animals, and the farmer.

The theosophist Rudolph Steiner made this point in 1924, in his lecture series on agriculture in Silesia. Steiner had been invited to that part of Austria, now in Poland, by farmers who were faced with severely degraded soils after years of plowing. Steiner said the problem was a lack of cosmic energy and that he could recommend some preparations to transfer terrestrial and cosmic forces back into the soil.

Many soil scientists describe Steiner's preparations as occult, geomantic, or superstitious, and, honestly, it is hard to prove one way or the other that you have harnessed cosmic forces. But it is equally difficult to deny that Steiner's prescriptions worked, and not just in Silesia. Just about everywhere they have been tried, Steiner's techniques, called "biodynamic agriculture," enhance soil aggregate stability, moderate soil pH, elevate available soil calcium, aid humus formation, and increase microbial and faunal biomass (earthworms and arthropods). With the significant increase in microbial diversity, there is a significant associated decrease in metabolic quotient, meaning a greater ability to use organic material for plant growth, and consequently healthier plants with higher nutritional value.

Despite Steiner's detractors, the process of creating soil *is* cosmic. The heat and cold of the changing seasons make soil. Thor's thunderstorms, arising from the spin of the Earth, the rays of the Sun, and the actions of wind on waves, affect the organisms in the soil and so contribute to making soil. Neptune's tides, stirring river deltas, peat bogs, and estuaries with the gravitational pull of the Moon and the planets, make soil. The pull of Jupiter that warps the Earth and opens cracks for volcanoes and earthquakes makes soil. The lunar cycles of 27.3 days pull tides in plants, humans, and animals alike. The Moon goddess Selene tells the bean when to root and reach for the sky, and when to drop its seed, fall back, and return into soil. Soil is inescapably cosmic.

The Soil Food Web

ONE HANDFUL OF GOOD GARDEN SOIL can contain more organisms than the number of human beings who have ever lived: 1 trillion bacteria, 10,000 protozoa, 10,000 nematodes, and 15 miles of fungi. If we creatures of flesh and bone were to suddenly vanish — whisked away by an alien transporter beam, say — the outline of our bodies would remain as a mat of nematodes over a core of bacteria — the part of the soil food web that covers our skin, inhabits our organs and helps our individual digestive and disease-preventative ecologies to function.

While relative densities and proportions vary among ecosystems, the same functional groups of organisms inhabit deserts, mountaintops, tundra, seashores, rainforests, and prairies. These functional groups include plants, bacteria, fungi, protozoa, root-feeding nematodes, fungal- and bacterial-feeding nematodes, nematode predators, arthropod shredders, arthropod predators, animals, worms, and birds.

Consider these as if they were players on a football team. The bacteria are the power backs — small bundles of protein with a high reservoir of nitrogen. Feed them and they can explode across the scrimmage line, knocking down anything in their path. The fungi are the wide receivers, fast and nimble. They extend immensely long arms of threadlike hyphae that can haul in carbon from one region, nitrogen from another.

Up until about one billion years ago, these two players were nearly the whole team, and plants hadn't arrived. Plants were the quarterback who took to the field and started directing the play. Plants put the fungi and bacteria to work as a team and began grinding up yards of turf. The bacteria take a handoff of nutrients from the plants — 20 percent or more of the photosynthetic energy coming to the plant goes into the soil as sugars, carbohydrates, and proteins exuded by plant roots — and use that to glue small clumps of soil together into larger aggregates (the offensive line). In this way, the one-two offensive punch of bacteria and fungi create a rhizosphere around the plant — a pocket where the quarterback can feel safe — with interior linemen and linebackers fighting off attackers such as parasitic nematodes, shredding arthropods, and root-rot fungi.

The early part of the game is a running game. Bacteria take over in early-succession communities such as bare earth, weeds, and vegetables. By halftime, in flowerbeds and row crops, fungi and bacteria are in equal balance. As the game clock winds down, the quarterback takes to the air, and bacteria are benched. In late-succession communities (shrubs and trees), fungi rule.

If the microbial organisms are the football team, and the gardener is the coach, earthworms are the engineers and architects of the stadium. They improve the playing and spectator conditions by creating air passages and corridors with their burrowing. Microarthropods are ushers and managers for the less mobile smaller organisms, helping them spread throughout the soil and onto the leaf surfaces. They get bacteria to where the nutrients are. The concession stands are run by arthropods that shred dead plant parts, making food accessible to bacteria.

The waterboys are the nematodes. These microscopic roundworms feed on nitrogen-rich bacteria, fungi, plant roots, or other nematodes, and then excrete excess nitrogen to the soil in a form that plants can drink, like Gatorade.

Knowing the bacterial/fungal ratio for the part of the game you are playing, you can coach different practices to encourage the best outcome. Tilling, hoeing, or digging up the soil, for example, favor bacteria over fungi. That is good for plant starts; most farmland and gardens are bacteria-dominated. To build soil among trees, encourage fungi with shredded leaf litter and wood chip mulch. To get more fungi, add compost that is formulated for the plants that are being grown, such as a fungal-inoculated biochar compost around fruit trees and conifers.

Removing grass from orchards is also a good idea, because grass is very bacteria-dominated. "Orchard grass" should be an oxymoron.

As a university researcher, Elaine Ingham did critical foundation work that underlies much of what we know today about the workings of soil ecosystems. When she first started studying fungi, they were considered to be nothing but trouble. Root-feeding nematodes — costing farmers tens of billions of dollars in crop yields each year — were the only kind of nematodes ever considered, without understanding that over-tilling, overuse of toxic chemicals, and overuse of inorganic salts (otherwise known as fertilizers) destroyed the healthy microbes that normally keep the "bad guys" in check. In healthy soils, bacterial-feeding, fungal-feeding, and predatory nematodes exist in high numbers. Ingham proved that you need the good guys to control the bad guys; get rid of everyone, and your plants will die.

Ingham came to unwelcome prominence when she intervened to stop her university's research into a genetically engineered, for-profit version of the *Klebsiella planticola* bacterium that lives on plant roots. She demonstrated that the new GMO bacterium, redesigned to make ethanol from cellulose, would have easily outcompeted the parent bacterium — and potentially turned all the world's plants' roots to mush, leaving us orbiting the Sun atop an orb of bubbling grey alcohol.

This did not endear her to her university, which banked on generous biotech industry grants, so in 1996, Ingham founded Soil Foodweb, a network of laboratories and consultants. Now with a dozen offices around the world, Soil Foodweb's mission is to work with, not against, soil life. Her essential tool kit includes a penetrometer, a soil corer, some sampling jars, and a microscope.

Out in the field, she takes soil samples, looks at root depth, and tests for where compaction layers might exist. Pressing down on the penetrometer at a rate of approximately 1 inch per second gives her a pressure readout. With healthy soil tilth, she can drive it to 4 feet deep with little effort.

If the penetrometer needle crosses 300 pounds per square inch (psi) on the gauge between its handles, she records the depth. That is the top of the compacted zone. She continues pressing down. Eventually, the needle falls below 300, and she knows she is below the compaction zone. If it never falls, the compaction zone is thicker than she can measure with that instrument. The penetrometer is designed to mimic a plant root. Root penetration decreases with compacted soil resistance, until almost no roots can penetrate soil with a resistance above 300 psi. Compaction can vary through a year; moist soil will be easier to push the penetrometer through, but compaction zones discovered when soil is moist will become much worse when the soil dries.

She walks to the areas of wheel traffic, takes transects in and out of the track, and records them separately. If there are deeply plowed areas in the field, she will measure penetration resistance in and out of the subsoiled zone. If there are planted rows, she'll take measurements in and between the rows. She wants to see the difference between trafficked and non-trafficked areas. Microorganisms in the soil have a hard time living in or moving through compacted zones, and oxygen can't move into compacted places, so anywhere compaction exists, there tends to be a predominance of harmful (anaerobic) bacteria, fungi, protozoa, and root-feeding nematodes, and a deficit of the beneficial (air-breathing) types.

- The four possible treatments for compaction are avoidance, alleviation, controlled traffic, and acceptance:
- Avoidance is the preventive approach: stay off the field until it can heal itself. Tractors that weigh less than 10 tons per axle usually cause compaction in the top 6 to 8 inches, but combines and grain carts that weigh much more create compaction as deep as 3 feet. At those depths, only reforestation will repair the damage.
- Alleviation at shallower depths can normally be accomplished with keyline, chisel, or subsoil plowing.
- Corn and soybean farmers who use global positioning systems for ridge-till, strip-till, or no-till can confine traffic between certain rows and avoid compacting the whole area. This practice is known as "controlled traffic" — lining up the combines, grain carts, and manure-handling equipment to confine the compaction to the same lines between rows (as illustrated on the cover of this book).
- Many farmers simply accept compaction as a trade-off for using heavy labor-saving equipment.

After Ingham has gathered her soil samples, she makes her way back to the lab, where she mixes the soil with water and puts a drop on a microscope slide. The penetrometer may not measure the first stages of recovery, so assessment of soil biology is also critical. Where soils have been degraded, we need to diagnose compaction, but then we need to introduce living organisms to treat and remediate the problem.

Freezing and thawing, wetting and drying, micro- and macro-aggregate building by bacteria and fungi, microarthropod and earthworm burrowing, and organic matter all build structure in the soil. Plant roots will find and grow through the spaces built by beneficial life in the soil. Inorganic fertilizers are not needed if healthy soil life is present in full measure. We can dispense with nitrate fertilizers (because they select for the disease-causing bacteria and fungi) and compacting equipment (because that reduces oxygen and helps root-rot fungi).

Given the knowledge Ingham has gained through years of looking at organisms in the soil, she is teaching growers how to do their own laboratory work with inexpensive microscopes. Using shadowing methods with simple light microscopes, growers can see many distinct kinds of bacteria, fungi, protozoa, nematodes, and larger soil critters. There are billions of individual bacteria, millions of fungi, hundreds of thousands of protozoa, and hundreds of good nematodes in a teaspoon of soil. And less than a tenth of the species have been identified.

It is the interactions of all these organisms that release plant-available nutrients, and most of that action is concentrated in the root zone. Trust Gaia to figure out how to do things economically, efficiently, and without loss of nutrients. As long as we can mimic her, we can easily get these benefits, too. It is the interactions of life in the soil that nourish us all.

Most of our soil "buddies" are decomposers that live on root exudates and plant litter — simple carbon compounds. Actinomycetes (more properly called actinobacteria, because they are really bacteria, not fungi) arrive early and are responsible for some of the "earthy" smells of freshly tilled soil. Tillage encourages these early bacterial species, while preventing beneficial fungi from extending their long threads and strands. Fungi act like a communications hub, carrying nutrients from decomposition sites to nearby root hairs.

The nitrogen-fixing bacteria genus *rhizobium* forms symbiotic relationships with legumes such as alfalfa, soybeans, edible beans, peas, and clover. *Rhizobia* colonize the roots of the host plant and convert atmospheric nitrogen (N_2) into plant-available amino nitrogen (NH_2). The bacteria supply protein to the plant, and the plant uses that protein for photosynthesis. In return, the host plant supplies the *rhizobia* with simple carbohydrates. A plant can donate up to 20 percent of its supply of carbohydrates, and nitrogen-fixing bacteria will then supply all its nitrogen.

When legumes are harvested, the entire root system, and quite a bit of plant residue material, is left in the field. If that plant material decomposes properly (without compaction, waterlogging, or other anaerobic conditions), the nitrogen in the left-behind plant material carries over to the next crop. This results in a "nitrogen credit," a gift to the next crop from the previous season's legume roots, leaves, and microbial community.

In a teaspoon of healthy soil, there are 10,000 to a million protozoa. These one-celled organisms are highly mobile, and they feed on bacteria and on each other. Because protozoa require 5 to 10 times less nitrogen than bacteria, nitrogen is released when a protozoan eats a bacterium. That released nitrogen is then available for plants to use.

Under the microscope, in a single field of view, Ingham wants to see hundreds of beneficial bacteria of 10 to 20 different types; a strand of beneficial fungi; and a few protozoa. Within 20 to 50 fields of view, she expects to find at least a few beneficial nematodes, which eat bacteria, fungi, and other nematodes. Nematodes need even less nitrogen than protozoa: between 10 and 100 times less than a bacterium contains, and between 5 and 50 times less than fungal hyphae contain. So when bacterial- or fungal-feeding nematodes eat bacteria or fungi, more nitrogen becomes available for plant growth.

Ingham's microscope may also reveal some potential soil menaces, such as nematodes that eat plant roots, or "switchers" that can switch from feeding on fungi to feeding on roots. These can be controlled by fungi that form impassable blocks, preventing root-feeding pests from finding the root; by nematode-eating fungi or arthropods; or by bacterial species that produce materials repugnant to the root-feeders.

If the numbers of bacteria, fungi, protozoa, nematodes, and arthropods are lower than they should be for a particular soil type, the soil suffers indigestion. Decomposition will be low, and valuable nutrients will be lost to groundwater or through erosion.

If the beneficials have been destroyed — e.g., through over-tillage, toxic chemicals, inorganic fertilizers, or too much lime or gypsum — we need not despair. Where do beneficial organisms come from? Compost. Elaine Ingham told me:

> The best way to manage for a healthy microbial ecosystem is to routinely apply organic material that is fully aerobic. No compacted, stinky, smelly black stuff. To keep garden soil healthy, the amount of organic matter added must be equal to what the bacteria and fungi use each year.
>
> Over the last 50–60 years, the attitude has been to get rid of the bad guys through pesticides and slice-and-dice tillage, not understanding that if you destroy the bad guys, you also get rid of the good guys. When we nuke soils and destroy life, what comes back are the bad guys . . . . Put your workforce back into place. They don't need holidays. Just make sure they're in your soil and feed them. Our job is to make sure there is a diversity of microorganisms, so plants can choose which organisms they need.

18

The Role of Ruminants

BEFORE THE FIRST BIPEDAL HOMINIDS appeared in North America, the land was cultivated by large quadruped mammals, typically leaf- and grass-eaters such as bison, deer, and moose. By the time the most recent major Ice Age allowed peoples to cross from Asia, the continent had gradually become cooler and dryer, and as it did, forests receded and grasslands expanded. Those grasslands, along with the pampas, savannas, and tundra, are part of the Earth's carbon store — about one-third of it, by recent estimates.

Grazing is necessary to maintain most rangeland soils, and since the wild ungulates that co-evolved with the native grasses have either been domesticated or gone extinct from most grasslands, their domestic counterparts — cattle, goats, sheep, buffaloes, and camels — will have to suffice.

Grasses do not have any mechanism to prune stems and dead leaves, because they co-evolved with herds of grazing animals. Most perennial grasses have their growth points, or buds, at ground level, below grazing height but exposed to the sun. Grasses are as dependent on grazers as grazers are on grasses.

When a grazing animal browses ground cover, it removes photosynthetic surface area, so the plant sheds root mass, which becomes labile carbon in the soil. As the animal browses, it is also exhales carbon dioxide, which wafts over the plant and is taken in through pores to build biomass. When the animal excretes, some carbon is lost into the atmosphere as carbon dioxide and methane, but considerably more makes it into the soil, along with nitrogen, phosphorus, and, of course, the rich microfauna contained in the animal's intestine.

In tropical environments that alternate between seasonally humid and arid, or in temperate zones that alternate between hot and cold, billions of tons of vegetation die each year. The microorganism population needed to decompose the vegetation would also die off from the changed seasonal conditions, were it not for the guts of large grazing animals, which remain a moist and fertile home for microorganisms. Over millions of years, a symbiotic relationship between plants, large animals, and microorganisms has evolved. Even in the harshest part of the year, microbes are able to break down the tough plant material, nourish the grazing animal, and sustain the life of the soil.

Over those millions of years, many species of large herbivores have come and gone, and have usually managed to keep up with the billions of tons of vegetation dying every year. Besides the microfauna in their intestines, the herbivores provide a reseeding service — hooves dig up the soil, digestion stratifies the seed, excrement buries it, rooting mulches, and urine fertilizes — all to ensure the spread and survival of the forage species. Grazing animals garden.

The dung of grass-eating mammals is home to a fascinating zygomycete fungus known as *Pilobolus,* which feeds on nutrients that have passed through the animal. Eventually it sends up tubes, each tipped with a fluid-filled bulb. On top of the bulb is a black spore packet. The fluid-filled bulbs act as water cannons, propelling the spore packets up to six feet away from the dung.

Pilobolus spores need to be eaten by an animal, which passes them through its digestive system, then excretes them with its dung, providing food for the fungus to continue. If the spores didn't get propelled away from the dung that *Pilobolus* lives in, they would never make it back to the animal, because most grazing animals will not graze on or near their own dung.

This dispersal strategy also serves other organisms, in particular, parasitic nematode worms that live in the gut of a grazing animal and send their eggs out with the dung. When the baby nematodes hatch, they climb up the *Pilobolus* tubes to the spore packet where they, too, are shot out of the cannon like a microbial amusement ride.

Allan Savory is a wildlife biologist who, in the 1980s, took a job as a game ranger in what is today Zambia. He was dismayed by the need to continuously cull populations of large animals like elephants because of damage caused from overgrazing. Savory's observation was that the problem was not too many animals, but rather an unnatural pattern of repeated or sustained land inhabitation, brought about by fragmentation of the natural wildlife range areas. In nature, animals move in large herds for protection from predators like wolves and lions, pausing to graze an area thoroughly, then leaving it for an extended period and returning after plant growth has recovered.

Savory experimented with a practice of "high-density mob grazing," in which relatively large numbers of herding animals concentrate and move frequently. There is no recipe. Herders monitor how the ecosystem functions, what it is that both the animals and the soils need, and adjust rotational grazing to give the best results for animal nutrition, plant recovery, mineral cycling, maintaining ground cover, providing a drought reserve by rainfall infiltration, improving soil life and structure, and so on. Savory dubbed this evaluation process "holistic management."

Grazing is necessary for the health of most rangeland soils. Eliminating grazing completely would be as wrongheaded as condoning overgrazing. But that also presents a dilemma. It takes an acre of grass, 1.3 tons of grain, and 435,000 gallons of water to produce one beef steer in the United States. A grass-fed steer finished on feedlot corn will have consumed in his lifetime roughly 284 gallons of oil, and seven times that if fed on grain from birth to market. Western civilization, and increasingly the rest of the world, has become hooked on a meat-centered diet. While it might have sustained a population of one, or even two, billion people, it cannot hope to feed seven billion and counting. Using just the grains now fed to cattle, on the other hand, would sustain a human population of 11 billion, or more.

The South American and Southeast Asian rainforests are being decimated by conversion to grazing lands. The meat and dairy industry — inhumane factory confinement in the main — produces 1 billion tons of manure annually, which is fouling wetlands, polluting rivers, and creating anaerobic dead zones where rivers meet the sea. Animal manure is now 60 to 80 percent of the total toxic waste generated in the United States, and the numbers are similar in many other meat- and dairy-exporting countries.

The total cattle herd in the world is 1.3 billion head; their weight exceeds the weight of humans on Earth. They consume 70 percent of the clean water, 85 percent of soy, and 43 percent of cereal grains. Their contribution to climate change is greater than that from automobiles. And yet, we need grazing animals to restore soils so they can hold more carbon.

And that is the dilemma.

Compost

MIDWAY THROUGH THE FIRST CENTURY of the current era, the Roman naturalist Palladius referred to goose droppings being turned into a soil amendment called "laetamen." A little later, Pliny the Elder mentions a wheat-like grain from the foothills of the Alps, secale, being decomposed, *pro laetamine est* (as if goose droppings). These references are the earliest we have of making compost. Of course, peoples whose lives were still very close to their gardens would probably not waste a lot of writing describing the ordinary processes of composting, which every child took as a daily chore, almost from the time he or she was old enough to walk.

Nor would it be without some irony that Pliny, who devoted his time to understanding how natural fertility arises in soils, would perish in a grand act of soil remineralization in which silica, calcium, magnesium, phosphorus, and potassium — elements formed in distant supernovae long before the birth of our Sun — were pumped from deep within the crust of the Earth and sprayed as fertilizer over the land. The Grand Farmer who cared nothing for Pliny but much for the soil was called by Romans "Vesuvius."

The value of compost is at once obvious. We want to make sure that we decompose toxic garbage and plant residues to create fertile soil, and we want to retain the moisture and nutrients in the soil that plants require. If we have good soil structure, root infiltration, and oxygen diffusion, we can end our dependence on petrochemicals and grow the most nutritious, most delicious, most healthy food. Compost, and only compost, gives us that.

Unless we lived in the flood plain of the Nile or Euphrates or some other major transport vessel for silt, or excavated a bat cave for guano, compost is all we ever used to make fertilizer — prior to the discovery by Colonel Edwin Drake of a "rock oil" seep on Seneca tribal land in Pennsylvania in 1859, which led, step by step, to Fritz Haber and Carl Bosch alchemically transforming air to ammonium, and the ascendance of nitrogen fertilizers.

John Adams, second president of the United States, was walking along the Edgeware Road in London in 1786, when he observed that

there are on the Side of the Way, several heaps of Manure, an hundred Loads perhaps in each heap. I have carefully examined them and find them composed of Straw, and dung from the Stables and Streets of London, mud, Clay, or Marl, dug out of the Ditch, along the Hedge, and Turf, Sward cutt up, with Spades, hoes and shovels in the Road. . . . This may be good manure, but it is not equal to mine. . . .

Adams religiously brought to his farm in Braintree, Massachusetts, dung from Boston, ash from potash works, seaweed, and marsh mud, and added it to

what will be made in the Barn and Yard, by my Horses, Oxen, Cows, Hogs, &c, and by the Weeds, that will be carried in from the Gardens, and the Wash and Trash from the House.[1]

Compost is garbage on its way to becoming soil, or more specifically, humus, the organic component of soil — although, as much as we depend on humus to provide our daily bread, we really don't know much about it. We know it has a carbon-to-nitrogen ratio of 10:1 and a moderate pH, but apart from that, all we can really say is that no two humus molecules are the same.

The only explanation for its wet, fecund, nurturing quality is that humus, like a human, is a collection of organisms and organelles that work together as a living entity. Remove their microbial communities (5 to 15 percent by weight), and both humans and humus wither and die. We may not know what humus is, but we know what it does. Fortunately, we also know how it is created. It comes out of the compost pile.

In a compost pile, the right conditions allow beneficial organisms — air-breathing bacteria, fungi, nematodes, protozoa, and microarthropods — to outcompete their rivals. Non-beneficial organisms typically grow better in anaerobic (oxygen-starved) conditions, so the first thing a good farmer like John Adams did was to make sure his pile had plenty of air. He mixed heavy, wet layers (dung, fresh vegetable wastes, marsh peat) with lighter, drier, and structural layers (charcoal and ash, sawdust, shredded leaves and twigs, straw).

To inoculate compost with beneficial organisms, Elaine Ingham suggests finding an undisturbed forest and taking small samples of the rich and healthy loam there. While local forest microfauna will be somewhat different from local grassland microfauna, studies have shown that the microfauna in forest systems includes much of the grassland microfauna. The prairie life may be hibernating, but they are there, waiting for the grassland diet to come along and wake them up like the smell of coffee brewing. In the conditions of a well-balanced compost pile, all of the microbiota will be represented.

When adding material to the compost, pace matters. If too much sugary or high protein food goes into the pile at one time, the bacterial feeding frenzy will consume oxygen and kill off beneficial organisms. Brief periods or clumps of anaerobic conditions may not be harmful and may actually increase diversity, but too much is detrimental. Once a pile goes anaerobic, beneficial fungi die or go dormant; nematodes, protozoa, and microarthropods die; nutrient cycling stops. In anaerobic decomposition, the decay of microbes and their nutrients produces volatile gases, and the result is a putrid, sour smell, similar to vinegar, rotten eggs, or ammonia. An early way to detect this process, before the smell, is to take pH readings, because acidity increases as the pile goes rancid. Never put putrid compost on plants.

Bacterial activity produces heat calories. The hotter a compost pile is, the better, up to about 74 °C (165 °F); more than that is too hot, and the pile should be turned or less manure used.

A typical fungal compost recipe is 10 percent manure, 40 percent green material, and 50 percent woody material. If the material is chopped too finely, however, undesirable anaerobic bacteria may come to dominate. A good compost pile stays dry and breathes. Adding biochar is an excellent way to accomplish this.

Biochar fresh from the kiln is a blank slate — an empty cupboard of cavities and micropores waiting to be filled by its microbial tenants. Placed directly in garden soil, sterile biochar may take a year or more to fully "charge" with nutrients, and it will draw in and store those at the expense of any nearby garden plants or field crops, meaning that the plants will likely grow poorly while that process is underway. If a gardener continuously adds sterile biochar, the garden could continue to languish indefinitely. If, instead, biochar moves from kiln or garden-store packaging to the compost pile, and then to the garden, it arrives in the soil fully charged and ready to create its "coral reef" effect. Nearby plants will thrive almost immediately.

A Tennessee company called CharBiological is currently trial-marketing two "charged biochar" products, one for large-scale landscaping, agriculture, and forestry, the other for potted plants and home gardens. Their "full spectrum" biochar contains 9 species of endomycorrhizal fungi,[2] 11 species of ectomycorrhizal fungi,[3] 15 species of bacteria,[4] 2 types of trichoderma,[5] humic acid, kelp and vitamins.

While it is often helpful to add nutrients to a garden to replace those consumed by the plants, in many cases the most important nutrients are made directly in the field by the soil community. Good compost resupplies food for a healthy soil ecosystem. Good compost tea, which we'll look at more closely in the next chapter, adds organisms; compost tea avoids the labor of spreading several tons of compost to the acre, and instead spreads beneficial microorganisms with a little bit of organic matter, and lets those microbes create the foods plants need.

Michio Okada's system of "Natural Agriculture," founded in Japan between the two world wars, attempts to foster a growing system as close as possible to what nature would do on its own. Okada's organization, Shumei, with 300,000 members worldwide, tries to farm without any additives. Beneficial insects, *bacillus thuringiensis,* sulfur, vinegar, and even manure are avoided because they do not exist naturally in the concentrations organic farmers typically apply.[6] Okada's philosophy was that nature already has everything it needs to thrive. The farmer's job is to optimize conditions and remove obstacles.

Rudolf Steiner also believed in minimizing inputs and letting nature do what nature does best. In the series of lectures in Silesia that later became the book *Agriculture* [7] and the foundation of biodynamic farming, Steiner prescribed eight different preparations for fertilizers, numbered 500 through 507, and gave details of how these were to be prepared. The first two are used for preparing fields, and the other six are used for making compost.

Preparation 500, for instance, is a humus mixture prepared by stuffing cow manure into the horn of a cow or ram, burying it into the ground (15–24 inches below the surface) in the autumn and letting it decompose over winter. In the spring, the contents of the horn are added to 10 to 15 gallons of water and whirled in different directions every second minute for an hour. In many German biodynamic farms, there are rainwater barrels or troughs at the corners of barns and stables that have stirrers hanging on chains, waiting to be used for this purpose. About two horns are needed for each acre of soil; in a large biodynamic farm, many gallons of fresh preparations are needed each year.

At first glance, it seems totally incredible that a biodynamic farmer can walk through his field with a whisk and a pail, spreading a mist of compost tea after the fashion of the chalky-white shamans Orellana saw waving censers at the riverside, and produce from that the most wonderful deep tilth and flavorful, relatively pest-free produce. And yet, one can witness this now on six continents, and it has been going on quietly, often pushing back against the opposition of government and industry, for nearly a century.

Recent scientific inquiries, such as those of Elaine Ingham, have tried to understand precisely *how* the preparations work. Ingham discovered that the whirling of the water barrel was significant. If the process remains aerobic, the complex set of organisms from the soil that colonized the cow manure and drew calcium and other nutrients from the horn will be transferred to the tea. That complex set of organisms is what is needed to establish a healthy soil food web. If the tea is applied year after year, the soil is improved and productivity increases.

20

Tea Craft and Designer Biochar

IF BIOCHAR IS STEEPED IN COMPOST TEA for a day or more, letting the microbes move into the pores and take up residency, and the biochar is then applied to garden, field, or forest, the fungal mycelium in the tea will immediately extend outwards from the biochar "reef" to the surrounding soil and draw in fresh nutrients, allowing a population explosion of beneficial organisms.

Starting with good compost, brewing good compost tea is not difficult, but it does require technique. Most beneficial organisms breathe air, so one of the goals of tea-making is to increase the flow of oxygen; this is usually accomplished by an electric bubbling aerator (unless we want to stir vigorously for some hours). As the tea is agitated and a dollop of freshly made compost is added, the population of the organisms multiplies and increases the demand for more oxygen. Optimal oxygen levels are 6 to 8 parts per million, or 70 to 90 percent dissolved oxygen in solution, which is the same as in the atmosphere.

Compost teas can be tailored to the needs of the soils and the crops being grown. For instance, in areas that have been overdosed for years with methyl bromide to kill *Phytophthora infestans,* the water mold that caused the Irish potato famine, application of 15 to 20 gallons per acre of tea supplemented with fish, kelp, or other micronutrients can restore soil health. Also, seed potatoes can be sprayed with a tea high in beneficial mycorrhizal fungal spores.

Compost teas can inspire "designer biochar" — formulations that have been tricked out to help particular crops in particular places. Adding nutrients to compost tea and steeping biochar in specialized teas will increase the population of beneficial organisms more rapidly, and also encourage the growth of particular organisms that may be deficient in a given soil. A biochar mix designed for blighted potato

Under the Microscope

The easiest way to tell how long it takes for a tea to reach its limits is to measure organism numbers over time. This is accomplished by taking samples and doing population estimates with the aid of a 10–40x bright field microscope with an Abbe condenser and preferably an iris diaphragm to shadow-outline the organisms. To be truly quantitative, the microscope needs to have differential interference contrast or epifluorescence attachments.

Fig. 11: *Binocular condenser microscope.*

A "field" is the circle of light that we see when we look through a microscope. Population estimation involves looking at 10 to 20 fields (moving the slide to bring each new field into view) and averaging. In a well-made batch of tea, there are more than 500 bacteria, a strand of fungal hypha, and a couple of protozoa, per field, and the occasional nematode to be discovered in some of the fields. All the organisms except the nematodes multiply in the tea-brewing process. Nematodes require several weeks to a year to go through a reproductive cycle, so the compost has to have a high population of good nematodes to see any nematodes in the finished tea.

It is a good idea to sample the compost going into the tea by mixing a teaspoon of compost into 10 or 100 teaspoons of water, and looking at it under the microscope. This provides an indication of how well the compost pile is performing and might also identify disease-producing pathogens that should not be seeded into teas.[1]

Good compost should have 15–30 micrograms of active bacteria per gram (µg/g) and 150–300 µg/g total bacteria. For bacterial-dominated compost, the active fungi should be 2–10 µg/g, and total fungi 150–200 µg/g dry weight. For fungal-dominated fields would be pre-colonized with the beneficial fungi and steeped in the kelp-laced tea just before being harrowed into the field.

To shift soils to be more alkaline, as for vegetables, grains, and grasses, we could use bacterial compost teas made by adding molasses, plant juice, or an algal extract. To encourage slightly more acid soils, as

compost, the active fungi should be 30–100 µg/g and total fungi 300–1000 µg/g dry weight.

Using the population estimate approach, those quantitative values translate into the following:

- For bacterial tea: 800–1000 bacteria per field, a strand of fungus in each two to five fields, several protozoa (preferably flagellates or amoebae, no ciliates) in each field, one or two bacterial-feeding nematodes in 20 fields.

- For fungal tea: 500–800 bacteria per field, one to two strands of tan, golden, or brown-colored strands of fungi per field, several flagellates or amoebae per field, and one bacterial-feeding nematode and fungal-feeding nematode per 20 fields.

Ten or more different types of bacteria should be distinguishable, for example, tiny round bacteria, medium-width round bacteria, short skinny rods, short plump rods, slightly longer skinny and plump rods, and chained and longer rods. Wide-diameter, colored hyphae are much more likely to be beneficial fungi than skinny, clear strands of fungi. If all that is present in the tea are the skinny fungi, get worried, and work harder to find some healthy soil to get the beneficial fungi to grow in your compost. If there are no fungal strands anywhere in your compost or tea, nature is try-ing to send you a message that a lack of oxygen enabled bad bacteria to take out all your beneficial fungi. If you see any predatory nematodes in the samples, count that compost and tea as being extra beneficial!

In a poor batch, there are fewer than 500 bacteria per field, usually in patchy clumps, no fungal hyphae, and no protozoa. If there are fewer than 25 bacteria per field, the tea is bad and should be thrown away (or put it on some weeds and see if it will stunt their growth).

Once the tea has achieved its maximum potency, it is important to get it to the field as quickly as possible, preferably within a few hours. If the bubbler is left on, teas can be kept active for three to five days before the microbes begin to go dormant or die.

for tree and root crops, we could add a little humic acid, a saponin, or ground leaves to make a fungal-dominated tea.

The limiting factors for growth of soil organisms in a tea-making apparatus or rain barrel are the size of the container and the available food. Once those limits are reached, it is time to "harvest" the live organisms by getting the tea into the garden, field, or forest before the beneficials begin to die or go dormant. Because tea-maker apparatus varies, batch time could take anywhere from 24 hours (in a 5-gallon bucket) to several weeks (in a 1000-gallon tank).

Mixing biochar with compost high in woody or fungal foods would best stimulate the soils below tree crops. Understory plants such as vines and berries need to be fungal-dominated, while row crops, vegetable gardens, and grasslands need bacteria to be dominant. Formulations of biochar compost or biochar soaked in compost teas can be designed to meet these different demands.

Soil organisms, like soil types, tend to be soil- and climate-specific. An organism that thrives in a glacial piedmont is unlikely to do as well in a sandy coastal plain or a field cleared from tropical forest.

Teas and biochars are not, by themselves, capable of immediately replacing the store of nutrients in a damaged field or forest, particularly any that have been damaged by years of chemical fertilizers and herbicides. Taking time to rebuild a field through applications of humic acids, compost, cover crops, and mulch assures that the soil food web will more quickly take root when it is seeded.

Reforming the soil is not difficult. We need to get oxygen 4 to 6 inches down, such as through keyline plowing. If pH is below 5.5, the soil is anaerobic or has been anaerobic in the past. Before you can improve pH, compaction has to be alleviated to allow beneficial organisms to grow. In fields where compaction is a problem, adding worms and worm castings to biochar and then subsoiling before injecting the char would be a way to re-oxygenate the compacted zones.

A single gram of well-made aerobic compost contains 16,000 µg nitrogen, 23,000 µg phosphorus, sulfur, calcium, potassium, and many other elements. If 50 to 100 pounds of nitrogen are being taken away each year in the form of crops and residues, then replacing that will require the application of 2 to 4 tons of compost per acre in the off-season. An easier process is to recover the missing minerals by drawing them out of the insoluble portions of the soil by biological activity. This is where compost tea comes in. Applications of compost tea with humic acid can replace much of the work involved with adding tons of compost.

In a month or two, with restored biology, the soil can recover to the point that plant roots can push more deeply. Where damage is deeper or the soil has been poisoned, recovery can take longer and repeated treatments may be needed. By regularly applying compost tea, more and more of the species of fungi, bacteria, protozoa, and nematodes will survive, establish, and start to remedy the problems, breaking down and sequestering even persistent pesticides, heavy metals, and salts.

Dilution of compost tea is not harmful as long as the water is the same temperature as the tea, which is usually air temperature, and is not chlorinated or otherwise treated to be antibacterial. If water is added to a sprayer, remember that the water doesn't count in what you are applying; it is the amount of tea per acre that counts. If run through sprayers or sprinklers, care should be taken to filter in order to avoid clogging with suspended particulates or sediments in the tea. Fine mists are not recommended for mid-day, because the droplet size will not protect the organisms from ultraviolet light.

Tea can be applied to foliage as a foliar spray at rates of 5 to 10 gallons per acre. On soils, it should drench at a slightly heavier rate of 15 gallons per acre. It should not be applied during heavy rain, but a light misting rain is not harmful and may even help. In temperate areas, it can be applied from the start of the growing season until first frost, and in tropical regions it can be applied whatever time of year plants most need protection.

Unlike good compost piles, which are hot enough to bake potatoes during the heat phase of the composting process, the temperature of compost tea should match the temperature of the soil, or leaf surfaces in the case of foliar sprays.

In the garden, bacteria, fungi, protozoa, and nematodes in soil will begin to solubilize nutrients from the rocks, parent material, sand, silt, and clay, as well as organic matter. Once life is established in high enough abundance, additions of compost should not be needed unless disturbances occur that harm soil health — for instance, large numbers of animals in small pastures, annual tillage, or severe fire or flood. In those cases, compost may be needed to replace the nutrients and restart the fauna.

There is no case in which inorganic fertilizers can't be reduced by at least 50 percent immediately, once foods for the organisms are present in the soil, the organisms themselves have been added, and they have survived and begun to flourish.

In the next year, fertilizers can be dropped another 50 percent, and by year three, on the slow plan, all inorganic fertilizer can be done away with. It is neither necessary nor helpful to maintain chemical fertilizer dependency.

Indigenous Microorganisms

At the Global Village Institute for Appropriate Technology in Tennessee, Cliff Davis gathers the starter culture for his compost teas from a bamboo grove next to a chicken coop. On a morning in the spring of 2010, he is out in the grove, inserting splits of bamboo stakes filled with oatmeal into the ground.

"The oatmeal was what was thrown into today's kitchen compost, left over from breakfast, and it has some apples and raisins in there too, and that should help," he says. He splits a length of bamboo about one foot long down the middle, creating two troughs, and spoons in the oatmeal compost. After he has completed this task, he rejoins the two sections and wraps rubber bands around to hold them together. Using a wooden mallet, he pounds the stake into the ground in the bamboo grove.

"You can get your tea starter in the mail, and those are called 'Effective Microorganisms,' or 'EM.' The way I like to do it is to gather them from the immediate vicinity, so I call them 'Indigenous Microorganisms,' or 'IM.'"

Davis has been harvesting soil fauna in this way for a few years, and preparing teas from the compost he will remove from the bamboo stakes after a few days. "I especially like this bamboo grove, which draws nutrients from the poultry area," he says. "It is the right type of bacteria and fungi, and they are very vigorous."

21

From Biochar to Terra Preta

WIM SOMBROEK'S VISION WAS TO PRODUCE a "terra preta nova" (new dark earth) by recapturing the recipe that had been lost for 500 years. Although Sombroek died in 2003 at the age of 69, the research that he began continues.

The terra preta soils formed on the bluff over the Rio Negro, where Orellana floated downstream 560 years ago, have been in continuous cultivation without fertilizer for the past 40 years. The farmers have no need to fallow their fields, and if they rotate their crops at all, it is more for weed control than to restore fertility.

The area of the Amazon covered by these man-made soils is greater than the area of Great Britain, which has its own isolated examples of dark earths from Roman times, but on a much smaller scale.[1] Tulane University anthropologist William Balée has concluded that the Amazon rainforest — the world's largest tropical ecosystem — is a human artifact. Balée says that the tropical forest that emerged during the collapse phase of Amazonian civilization could have been the result of opportunistic tropical flora and fauna seizing the opening provided by extraordinary soil fertility. The centerpiece of that fertility is the biochar made by the lost civilizations.

Amazonian dark earths have up to 150 grams of hard carbon per kilogram of soil — 15 percent — in comparison to the surrounding soils with 2 to 3 percent.[2] Total carbon stored in these soils is 100 to 200 times that of adjacent soils.

The depths of organic matter go well below the 4 to 8 inches common in that region. Sixteen to 20 inches is average, and 80 inches is not unheard of.

Amazingly, terra preta soils seem to also have the capacity to grow themselves. William I. Woods explains the phenomenon as a consequence of the vibrant life of the soil. Soil-dwelling creatures that themselves contain carbon — some of it taken from the roots of nearby plants, some coming from farther away — die; their bodies decay on the biochar reef, and they add to the volume of carbon, extending the horizon of the dark earths. Mycorrhizal fungi also attach to the large charcoal particles. They fix additional carbon, stabilize the soil with glomalin, and increase nutrient transport to nearby plants, which provide roots that later shed and decompose to add still more carbon.

Biochar, which was called "agrichar" until an enterprising company in Australia trademarked that word, is the first ingredient in the terra preta recipe. The term "biochar" was coined by the late Peter Read of the

Pyrolysis

If you have ever watched wood burn in an open fire, you have seen pyrolysis. The process of combustion heats up the woody biomass, and the pyrolysis reaction begins. Long-chain carbon compounds such as cellulose and lignin are broken into shorter-chain carbon compounds such as phenols, aromatics, methane, and carbon dioxide. These pyrolysis gases combine with oxygen, react, and voilà! Fire.

If the volatile gases had not come in contact with oxygen, they could not have oxygenated, and there would have been no flame. This sometimes happens when you close the vents on a wood stove and leave it to simmer overnight.

Pyrolysis, the heating of the woody biomass and the release of gases, produces three products. First, there are volatile gases, and these are usually burned away. Second, there is a tarry, bubbly liquid called "wood vinegar," composed of a variety of different carbon compounds, and this is also typically burned away. Third, there is the carbon residue left behind — the char — which is about 80 percent pure carbon.

Pyrolysis requires heat, and typically that comes from the fire itself, but if oxygenation is allowed to proceed it will consume all of the char as well as the liquids and gases. What you will be left with is just ash, which is the usual case in an open wood

Centre for Energy Research in New Zealand to describe charcoal used as a soil amendment for agriculture. It is a fine-grained, highly porous charcoal produced by burning in an oxygen-starved environment.

Like crystalline clay, charred material is not electrically neutral along its exposed edges and outer surfaces. It attracts to itself oppositely charged particles such as calcium, nitrates, phosphorus, and silicates — the building blocks for organic life. The cavities of biochar become cupboards of these ingredients from which microbial cooks whip up entrées to be carried by fungal waiters to the hungry plants.

Carbon particles formed at low pyrolysis temperatures (around 300–500 °C) are made of a hard, poly-condensed aromatic grouping of carbon atoms that is recalcitrant and has very long chemical and biological stability. When pyrolyzed, only the portion of the carbon that was not already bound to oxygen is available to become fixed carbon. The rest is in the form of volatiles — tars — that may stick to the surfaces of the biochar at low temperature or be burned away at high temperatures. The tars contain oxygen, and as they paint the cavities

fire. The function of pyrolysis reactors, kilns, or retorts is to separate the fire from the material being heated and to burn as little of the gas and liquid as possible to achieve the heat required. The portion of the gas and liquid that is not burned can then be harvested and put to other uses.

The amount of char, gas, and liquid harvested from combustion depends on a number of factors, among them the time, pressure, temperature, and type of fuel. Typically, pyrolysis occurs in woody biomass between 200 and 600 °C. Reaction products vary with the conditions of the burn. At higher temperatures, higher concentrations of gas are produced. To get the most char, lower temperatures for long periods of time are optimal. Temperature and timing are also key to maximizing the liquid "wood vinegar," which is a sort of "bio-crude" and can be further refined into liquid fuels, preservatives, and other products.

Temperature and timing can be controlled to retain nitrogen, calcium, potassium, or other trace elements, or to otherwise improve the yield of biochar, the product that the soil food web will most appreciate.

in the char, they enlarge the reactive surface area. Would-be microbial residents are attracted to the many cabinets of the cupboard and to the paint used on those cabinets.

Good biochars, with high carbon content and good adsorption qualities, appear to be achieved at 500–700 °C with ample gas flow. There may be other benefits of oxygenated tars left on the char due to very low-temperature slow pyrolysis (under 500 °C), but these seem most likely to serve as short-term food for microbes.

Temperatures are also critical for the slow oxidation of biochar that creates carboxylic groups. These increase the cation (positively charged ion) exchange capacity in the soil, drawing nutrients to the reef.

Wood biochar, as opposed to a char made from grasses, corn stover, algae, or other forms of cellulose, has an internal layer of biological oil condensates that soil bacteria consume, with beneficial effects on microbial growth. Poultry litter also has this attribute. At high temperatures, these condensates are volatized and burned away.

Scientists like to call the hard form of carbon, whether tarry from low-temperature pyrolysis or tar-free from high-temperature or pressurized burns, "recalcitrant" because it does not want to bond with anything else and is quite happy keeping to itself.

Although clays, composts, and other factors may play a part, recalcitrant carbon is the backbone of the terra preta recipe. A study performed in Finland by researcher Janna Pietikäinen tested plant response using three different high-porosity materials — zeolite, activated carbon, and biochar. These tests showed — contrary to her expectations — that microbial growth was substantially improved only with biochar.

❧

Not all char is biochar. Burned crudely, in the kinds of low-cost, handmade kilns one can find in much of the developing world, forests will still be unsustainably harvested, and emissions of greenhouse gases will still rise. Burned to more exacting standards, in well-sealed kilns, with carefully managed curing times and temperatures, with characterization of the feedstocks and fuels, emissions can be greatly reduced, and the yield is a high-carbon biochar, ready to be "charged" with nutrients and added to the garden or farm.

An entire range of organic materials, including poultry litter and sewage sludge, can be charred, and the resulting biochar will have trace minerals and pH that reflect the differences. There is a legitimate environmental concern about potentially excessive levels of heavy metals in biochars that come from treatment of sewage or other waste streams. To ensure safe use of biochar in agriculture, it is very important that an organic standard be developed that takes into account both the sources and the production techniques.

One of the tests being recommended for biochar scientists by the International Biochar Initiative is the Worm Avoidance Test. Any biochar that offends an earthworm is probably not something you want to put into your garden. Other tests that are useful for gardeners and farmers acquiring a new source of commercial biochar are pH, TDS (total dissolved salts), and seed germination.[3]

Charred macadamia nut shells in Hawaii are a different product than charred pine sawdust in Canada or charred chicken litter in East Timor. Some portion of the original chemical composition of the feedstock remains, even after it has been subjected to a high-temperature reduction. Oils beneficial to soil bacteria may or may not be present. Pore surfaces vary. Labile-to-recalcitrant carbon ratios will be different, and acidity will vary.

Hugh McLaughlin, a soil scientist and engineer working on characterizing biochars for the International Biochar Initiative, recommends that every farmer adding biochar to soil keep a sample in case there is a need for later analytical testing. "Relatively dry biochar in a sealed jar will store at ambient conditions for years without significant change," says McLaughlin. "A pint jar of a biochar sample is absolutely indispensable in determining what went wrong, or right, with a given biochar trial."

Some biochars may have a high pH, and if applied to a soil that is already high in pH, the plants would suffer or even die. If applied to a low-pH soil, an alkaline biochar would be perfect as a liming agent, and plants would thrive.

Charring of feedstock with no inorganic components (purified cellulose and lignin) produces chars with acidic pH. Charring of switchgrass or pinewood at low temperatures produces a neutral or slightly acidic char. As the charring temperature is increased, the amount of acid in the char decreases, the amount of char decreases, and the ash content increases. This enables the alkalinity in the ash to overcome any acidic functional groups attached to the carbon, resulting in alkaline biochars. Since pinewood produces less ash than switchgrass, it takes higher charring temperatures to produce an alkaline char for pinewood than for the switchgrass.

When a rose grower marches into the garden shop and asks for biochar, or a farmer calls his local cooperative to purchase a truckload of biochar for delivery to his field, they don't want to hear that the product they are getting this time may be different from what they purchased last time.

That's where the umbrella of biochar advocacy groups — the International Biochar Initiative and its local chapters — comes in. Creation of the IBI was one of Wim Sombroek's final accomplishments. IBI is standardizing methods for characterizing feedstocks, analyzing end products, making life-cycle assessments of energy and

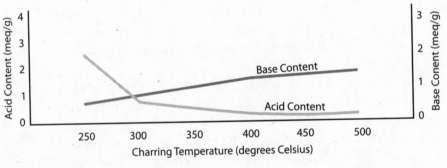

Fig. 12: *Acid/base content of switchgrass chars (after Rutherford, Rostad, and Warren[4]).*

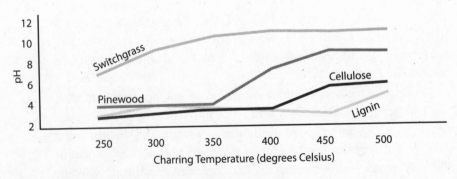

Fig. 13: *Acidity of various biochars (after Rutherford et al.).*

water use, and promoting best practices for biochar manufacture. It is like the group of farmers and gardeners who met together to set standards for the use of the word "organic" half a century ago. This time, the fate of the Earth could well hang in the balance.

There are other parts of the terra preta recipe that remain a mystery. Amazonian terra preta, for instance, has been found to contain:

- burnt clays
- human and animal excrements (rich in phosphorous and nitrogen)
- hunting, fishing, and cooking refuse, such as animal bones and tortoise shells (rich in phosphorous and calcium)
- ash residue from incomplete combustion (rich in carbon, calcium, magnesium, potassium, and phosphorous)
- the biomass of terrestrial plants (compost, rich in nitrogen and carbon), and
- the biomass of aquatic plants (reeds and algae, rich in calcium, magnesium, potassium, and phosphorous).

Field studies in Brazil have shown that not all terra preta is the same. Some could have been produced by a slash-and-char technique of open field burning. Others might contain more fish bones, composted vegetal material, or highly acidic dried leaves. The lifestyle of the village that produced the terra preta can be reconstructed from the soil they left behind, and not all villages had the same diets or foodscapes.

The peregrine earthworm *Pontoscolex corethrurus* might be the sous-chef that stirs the pot wherever terra preta is being cooked up. It is widespread in Amazonia and has been collected in clearings after burning processes. The worm, which has a high tolerance for poor, parched soils, has been shown to ingest pieces of charcoal and mix them in its gut, excreting a finely ground form that mineralizes the soil. Thin, regular layers of biochar are perfect places for *Pontoscolex corethrurus* to work its magic.[5]

Bruno Glaser, one of the leading soil scientists in this field, points out that many of the interactions between the char, the soil, and the microorganisms only come with time. In particular, the role of burnt clay in the recipe is not understood. Glaser believes that to make biochar behave like terra preta could require 50 or 100 years, as the microbial transformations of the substrate evolve into dark soil.

Others are not so certain. Research by two Australian scientists, Drs. K. Y. Chan and Z. H. Xu, suggests the nutritional quality is dependent on both the nature of the feedstocks and the pyrolysis conditions under which the char is produced. At relatively low charring temperatures, under 300 °C, any cellulose will produce a char, and water, carbon dioxide, and carbon monoxide will be released. Increase the temperature to 300–500 °C, and a rapid depolymerization occurs, producing "wood vinegar" or "bio-oil." At higher temperatures, 500–700 °C, the char is almost 90 percent carbon, and has a higher surface area, attributable to water vapor molecules creating very small (nanometer-size) pores that increase surface area by orders of magnitude. These changes, in turn, have important effects on the nutrient retention ability, cation exchange capacity, and anion exchange capacity. Other factors, such as the heating rate and particle size of the feedstocks, are also important.

While low temperatures may be desirable, there is no single best recipe for making biochar, and there is no single best recipe for making terra preta soils. In every location, the final product will vary, depending on the available plant sources, conditions, and methods. For this reason, it is important to be able to test both biochar and soils rather than relying on standard formulations to achieve the desired results.

Making Charcoal

IN MEXICO'S YUCATÁN PENINSULA, between Cancún and Merida, there is a little town called Esperanza — "Hope." The local people say it was called Hope because it used to be so far into the forest the people hoped that if they gave their town a name, one day a road would come to them. For as many generations as they can remember, the people of Esperanza have made and sold charcoal.

A road now passes through Esperanza, although it was not built for the charcoal-makers. As the Mayan Riviera sprawled outwards from Cancún, roads were needed to take tourists to the ruins at Chichén Itzá, the orchid *corchals* of Solferino, and whale shark cruises off Isla Holbox. Today buses full of tourists pass right through the center of Esperanza, and some tourists may notice the tall stacks of bundled charcoal in front of every home.

The people of Esperanza make their charcoal the same way people do in Africa, India, Peru, Indochina, or anywhere else there are remnants of ancient traditions. The Phoenicians used wood tars from charcoal as caulking for ships. Egyptians used the high-temperature properties of charcoal to make glass and to smelt copper, and used the wood vinegar from pyrolysis for dyes and embalming.

The Bronze Age arrived when blacksmiths learned to alloy tin and copper at temperatures that could only be obtained in charcoal forges. The Iron Age followed in much the same way. In some places, excessive felling of forests to fuel the iron, tar, or lime cement industries brought

Fig. 14:
Charcoal for sale in
Esperanza.

climate change and ruin. The Finnish forests of today date from a mere 300 years ago because that is when large-scale tar and charcoal production ended.

In many other parts of the world, charcoal production was managed more sustainably. Coppicing, for instance, involved cutting trees back periodically to stimulate growth and harvest firewood. Coppiced wood was of regular size, easy to handle and load, and easy to transport the short distance from woodlot to ironworks. The process was carbon-neutral, or even carbon-negative, because every time a tree was coppiced, some of its root mass died to compensate the loss of photosynthetic surface.

In the 18th century, the iron industry switched to coke — a metallurgical product of coal discovered by Abraham Darby — and both the protection of the woodlots and the carbon-neutral fuel balance were abandoned. Still, the Mayan inhabitants of Esperanza continued to make charcoal and to teach their children their method. They knew nothing about coal.

The Mayan method, and that of millions of other people the world over, consists of arranging lengths of wood on their ends to form a conical pile. Openings are left at the bottom to admit air, with a central shaft to serve as a flue. The whole pile is then covered with dirt and moistened. Firing is begun at the bottom of the flue and gradually spreads outwards and upwards. Holes at the base to release smoke are reduced until, in the oxygen-free interior, pyrolysis begins. After a process lasting about five days, the pile is quenched with water and

Fig. 15:
Mayan charcoal-making in Esperanza: The Mayan method of charcoal-making starts with arranging lengths of wood on their ends to form a conical pile.

Fig. 16:
Openings are left at the bottom to admit air; a central shaft serves as a flue.

Fig. 17:
The woodpile is covered with dirt, moistened, and fired.

uncovered. In Esperanza, as in most other places, 100 parts of wood yield about 60 parts of charcoal by volume, or 15–25 parts by weight. What is wrong with this traditional method is that it is both wasteful and dirty. A large part of the wood is consumed to dry and pyrolyze the charcoal. When the fire is lit, an immense volume of smoke, mostly carbon dioxide and monoxide, escapes to the atmosphere. During the week that the fire burns, smoke escapes, and when not being inhaled by the colliers managing the fire, it finds its way to the atmosphere. This is just about the worst kind of pollution there is — long-chain hydrocarbons, poly-aromatics, tars, ash, and black carbon soot. If it is not raining, a smoke haze lingers over the town.[1]

To compound the problem, most of the charcoal gets used as fuel for cooking stoves, grills, and braziers. Not only does nearly all of the carbon then find its way skyward, the stoves of the world's poor are typically very poorly designed — even just open fires. Collectively, they are responsible for an estimated 1.6 million deaths per year from smoke inhalation. [2]

The smoke of virtually any biomass source, for all its green roots, is a caustic chemical brew containing benzene, butadiene, dioxin, form-aldehyde, styrene, and methylene chloride. Every branch and leaf adds toxic compounds so corrosive they can riddle untreated steel. Lungs are easy pickings. The leading killer of children worldwide is not malnutrition, malaria, or endemic disease. It is pneumonia, an infection of the lungs.

The cooking fires for three billion people each produce about as much carbon dioxide as a car. A single gram of black soot warms the atmosphere as much as a 1500-watt space heater running for a week.

One of the simplest charcoal-making "stoves" is the Stump Burner — a pipe of rolled sheet metal with air holes at the bottom, placed over a stump, loaded with dry biomass, and set on fire. Combustion is controlled by partially blocking the upper end of the tube. In a refined version, a slightly larger tube is placed on the outside to preheat the air going in at the bottom. The top of the inner tube can have a metal cooking plate, and a stovepipe with damper to create and control the draft. It is simple, cheap, and portable, and reduces the need to dig up stumps. However, just like the Esperanza piles, this method is not clean-burning.

While making biochar as a soil amendment can be as simple as starting a fire in a row of corn stover and then covering it to smolder, that is no way to save the planet. Biochar needs to be ethically manufactured in a way that is safe, easy, and clean. We also need to think about those three billion people inhaling smoke from cooking fires.

Stove Wars

MANY PEOPLE HAVE BEEN GRAPPLING with the pollution problems associated with stoves, and some very good alternatives have been emerging. Biomass-burning stoves, which are now an environmental problem, may become one of the world's most promising environmental hopes.

If we separate the biomass from the burning flame, we can have a clean burn that reduces all of the volatiles to a little CO_2 and water vapor and gives us biochar instead of ashes. The trick is to engineer that separation. One simple way to do that is with a stove insert, which could be a metal cylinder or box that can be tightly sealed but has small holes to allow a controlled release of pyrolysis gases into the fire chamber of the stove. The insert is loaded with a biomass feedstock and placed into the fire; when it reaches pyrolysis temperatures, the gases escape and ignite, reducing the fuel requirement for the stove while producing biochar. After a time, the gas flames die, the insert can be retrieved, the biochar removed and quenched, and the process repeated.

One of the pioneers in low-tech biochar stoves is Folke Günther. Before experimenting with biochar, Günther worked eight years as a lecturer in human ecology at Lund University in Sweden, specializing in functional integration between human settlements and agriculture for recycling soil nutrients and saving energy. After experiments, Günther came up with his "Holon Stove," designed to produce biochar more cleanly and efficiently than traditional methods.

The Holon Stove is simply one metal drum filled with dry biomass inserted inside another metal drum. The inner drum has to be small enough that the airspace between the two drums can be filled with flammable kindling, which is ignited to heat the inner drum. The inner drum is sealed at top and bottom but has holes to release pyrolysis gases, which will catch fire, and these flames further heat the inner drum from without, continuing the process after the starting kindling is consumed.

Biochar enthusiast Kelpie Wilson developed a double-chamber model using a steel barrel for the inner biochar container and cement block for the outer initiating fire. The advantage of this design is that it is inexpensive and can make larger batches. The disadvantages are the same as for the Holon stove: thin steel burnout and long cooling time.

There is no reason why the inner drum has to be the pyrolysis chamber and the outer drum the starting fire. There are many builders who have chosen to reverse the two, charring biomass in the larger outer chamber from an initiating heat source at the center. The Anila Stove, developed by Professor U.N. Ravikumar of Mysore University, and now in commercial production in India, works from this principle.

Fig. 18: *A biochar-producing insert for a wood heater.*

Fig. 19: *Folke Günther demonstrates his Holon Stove.*

A scaled-up industrial version is the Torbed® reactor system developed by Polow Energy Systems BV in the Netherlands for gasification and flue-gas cleaning.

While two-chamber stoves such as these are very simple to construct and cleaner and more efficient than traditional charcoal-making, they still have some disadvantages. They take a long time to cool and reload, and since they are typically made of thin steel alloy, they may only last for about 30 burns before burning through the metal. The Anila is cast from aluminum to avoid the burnout problem, but aluminum, which is expensive to refine and smelt, brings its own set of problems.

Stillwater, Minnesota, engineer Sean K. Barry included a layer of insulating foam inside the burn chamber to extend the life of the steel and improve heat retention in the inner chamber. His version of the double-barrel kiln is among those being tested by the United Nations Food and Agriculture Organization with the goal of developing an inexpensive retort that could be used by two people to make 700 to 1000 pounds of biochar per day.

Tom Reed, a professor of Chemical Engineering at the Colorado School of Mines, has been working on wood-gas stoves since a visit to Africa in 1985. Reed applied gasification to the problems of wood cooking, eventually coming up with the Woodgas Camp Stove, which can easily be replicated from used paint cans or number-10 tins (the size used in institutional cooking). His company, Biogas Energy Foundation, sells two sizes of stoves, starting at $55. The stoves use two AA batteries to power a fan that forces pyrolysis gases to the combustion area.

Top-lit updraft (TLUD or "Tee-lud") stoves are one of the simplest ways to make a carbon-negative cookstove. Hugh McLaughlin's

Figs. 20 and 21: Hugh McLaughlin's "Toucan" stove, made from a used one-gallon paint can.

"Toucan" is one of the smallest and simplest TLUDs, requiring only two tin cans (hence "Toucan"), although a third can is recommended for a chimney. A used one-gallon paint can is perforated at the bottom with a can opener to form the outer shell. A number-10 can is cut down to about two inches and perforated as a "crown." The optional chimney comes from a 44-ounce frozen juice container.

A scaled-up version of the TLUD has been constructed by farmer Dick Gallien of Winona, Minnesota, from a railroad tank car and bunker silo materials. The large tank is designed to make biochar in the winter, with enough heat generated from the charring process to heat his farm buildings.

Another early stove developer is Christoph Adam, an international aid specialist who constructed his first "Adam-retort" at the Auroville ecovillage in India, using $700 in materials and two local brick masons.[1] In the Adam-retort, about 3 cubic meters of biomass (in the case of Auroville, a half-ton of coconut shells) is converted to 500 to 750 pounds of charcoal per 30-hour batch. Adam calculates that the initial investment costs can be paid back in about three months from sale of the charcoal. Where charcoal costs are high, the payback is faster.

Peter Hirst, who started the company New England Biochar, purchased the US rights to make and sell Adam-retorts and constructed the first all-steel version designed to be shipped from the factory. The chambers are 4 x 4 x 8 feet; with appendages it stands 10½ feet long with two 8-foot stacks. Operated by two people, it can make 700 to 1000 pounds of biochar per day. After trying one batch, Hirst reported:

> Once pyrolysis kicked in, things really got spectacular. We tuned for a perfectly clean burn, even with a massive excess of producer

Fig. 22: The Adam-retort.

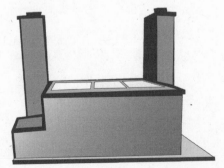

gas, and during the course of the test developed significant design advancements for energy capture, emissions reduction, and phase-one drying acceleration. There is not only excellent pyrolysis with this unit, there is serious excess energy available for doing all kinds of things. The unit burns completely clean when kept in adjustment and for over 2 hours of the 4½ hour pyrolysis, the unit maintained a solid flame throughout a combustion path 24 feet long with no visible emissions. Zero smoke from the stack for all but a few minutes of the 4½ hours. None. We expect to unload about 700 pounds of char tomorrow morning and go after it again with air-dried limbwood.

Moving up from backyard and village-scale experiments to commercial production scale, the field is getting crowded quickly. Here were a few of the contenders at the start of 2010:

Biochar Engineering Company (BEC) developed a mobile biochar production reactor that sits on a skid or wheeled trailer. It can be taken to sources of waste woody biomass, such as forests blighted with beetles, hurricane cleanup areas, and sawmills with waste material that would otherwise be left to decay or burn. The BEC reactor converts 1000 pounds of woody biomass per hour into 250 pounds of biochar. The reactor costs $100,000, and the first production units were immediately sold to the US Forest Service. BEC is experimenting with peripheral devices to produce methanol and dimethyl ether for liquid fuels, along with process heat, steam power, and hot water — all from the same mobile reactor. By producing standardized, modular units that are easy to install, or to remove and resell, BEC is eliminating some of the obstacles to small-project financing.

A similar business model of affordable, transportable pyrolysis plants, specifically for agricultural residues, forestry wastes, and transition crops, is being pursued by Advanced BioRefinery Inc. (ABRI), Alterna Biocarbon, and Agri-Therm, all of Canada; Renewable Oil International of Alabama, US; and Black is Green of Maleny, Queensland, Australia, sellers of BiGchar™.

The eGEN series developed by John Gelwicks in Redondo Beach, California, is a continuous-batch biochar producer that co-generates electricity. The CR-2 model turns 160 kilograms per hour (kg/hr) of waste poultry litter, almond shell and husk, or grape-crushing waste into 40 kg/hr of biochar and 64 kilowatts (kW) of power. The CR-3, on sale in 2010, will turn one ton of waste into 250 kg/hr of biochar and 400 kW of syngas, with some variation based on feedstocks.

BEST Energies Inc., of Madison, Wisconsin, and New South Wales, Australia, has developed a slow-pyrolysis, continuous-feed technology that converts biomass wastes — typically poultry litter, dairy manure, paper sludge, and nutshells — to biogas, pelletized fuel, and "Agrichar®."

A similar strategy of exploiting agricultural animal wastes is being pursued by Edward Someus with a Hungarian company called Terra Humana Clean Technology Ltd. After pyrolyzing the biomass, Terra Humana is developing methods of inoculating the biochar with a microbial carrier and formulated fermented nutrients.

Carbonscape, of Blenheim, New Zealand, has developed an electric microwave carbonization process that converts 40 to 50 percent of wood debris into charcoal; 1 ton of carbon dioxide can be fixed as charcoal each day. CleanFuels, of the Netherlands, is working in the other direction, building plants that turn pyrolysis gases into commercial electricity. Dynamotive, based in the US and Argentina, produces CQuest™ biochar and bio-oils for fuels. Pro-Natura International and Eco-Carbone, of France, are producing "green charcoal" in Brazil and Senegal from jatropha, savanna grasses, reeds, wheat and rice straw, cotton and corn stover, rice and coffee husks, and bamboo.

Bioenergy LLC, of St. Petersburg, Russia, has created the Polikor family of large-scale reactors. The first "Polikor-3" was put into operation in December 2002 in Arkhangelsk. Five kilns of this series are now in operation. A similar company is Bioware, in Brazil, which builds fast-pyrolysis reactors using bubbling fluid-bed technology to produce bio-oil and charcoal powder.

Biz-Solutions began in 2002 as a Canadian real estate company, evolving into a renewable/alternative energy company. Biz-Solutions plans 30 new pyrolyzation facilities in the Southwestern US by 2014.

Envipower, of Lyngby, Denmark, a maker of residential apartment building biomass boilers, has started adding biochar as a co-generation product of its heaters. It now makes two-stage units for combined heat and power (CHP) up to 5 megawatts.

Burning wood for electricity only, about three to four trees need to be burned to recover the energy contained in one. CHP plants with advanced wood combustion can recover up to 90 percent of the energy in the biomass. By adding the step of pyrolysis and gasification, they can push that efficiency even higher and co-produce biochar to sequester carbon and improve agriculture.

R & A Energy Solutions of Ridgeville, Ohio, provides modular CHP pyrolysis equipment for the dairy, cattle feedlot, recycling, waste-hauling, municipal utility, and auto-shredding industries. Its PyroGen™ systems are available in 250-pound, 500-pound, 1000-pound, 2000-pound, and 4000-pound-per-hour sizes. The larger scale (2 tons per hour) will produce up to 2.2 megawatts of power and Pyro-Oil™, Pyro-Char™ (unsuitable as a soil amendment), or Bio-Char™ (suitable for agriculture). PyroGen is currently experimenting with conversion of 2000 gallons per day of flushed swine manure in North Carolina.[2]

Renewable Oil Corporation (formerly Enecon) is developing a renewable energy business in Australia that will convert wood residues and green waste into electricity, heat, and chemicals. Its first commercial-scale pyrolysis plant will process 200 tons per day and provide approximately 8 megawatts of electricity. A similar business strategy is being followed by Sustainable Power Systems, makers of Vee-Go,™ Biochar Xtra, and Vertroleum®.

Eprida, which began operations in Athens, Georgia, in 2002, uses a hybrid pyrolysis/steam-oxygen gasification process that generates hydrogen (used to make gas-to-liquid biodiesel, ethanol, or dimethyl ether for fuels), ammonium bicarbonate (for fertilizer), and biochar. The Eprida process can be used to scrub emissions from power plants, cement factories, and steel mills to recover nitrogen and sulfur without expending additional energy.

Quadra Projects says its patented QES2000 system is "the most advanced pyrolysis and gasification system in the world, specifically designed to convert palm husks, used tires, plastic waste, and municipal waste to carbon black, biochar fertilizer, and/or fuel oil or fuel gas for the production of electrical power, without any measurable environmental pollution or ash to be landfilled." At the start of 2010, its stock (symbol: QPRJ) was trading at 10 cents per share.

Some biochar-related companies don't actually make any product; they help other companies locate feedstocks, buy and install systems, secure regulatory permits, operate the completed plant, and broker the biochar and power.

In the town of Morbach, Germany, Juwi Group has turned a 370-acre former US military munitions depot into an "energy landscape" with 14 wind turbines, solar parks, a biogas plant, and a wood pellet factory. In 2010, Juwi will begin selling its own "terra preta" as a formulated potting soil, using biochar from biogas production residues. "We have solved the puzzle," says the group's Joachim Böttcher. "We are the first and the only ones in the world who can now produce terra preta."

As we've seen, biochar needs to be ethically manufactured in a way that is safe, easy, and clean, and we need to remember the three billion people presently inhaling smoke from cooking fires.

Simple stoves designed to take on this challenge have been explored by N. Sai Bhaskar Reddy of the Geoecology Energy Organization of Andhra Pradesh, India, for many years. Reddy has even developed a lighting system using a wood-gas updraft stove.

In 1980, the rocket stove was conceived by two of the founders of the Aprovecho Institute, Larry Winiarski and Ianto Evans. It burns small-diameter fuel such as twigs or small branches and achieves high combustion efficiency through directed ventilation. The rocket stove's main components are a fuel magazine, a combustion chamber, a vertical chimney to provide the updraft needed to maintain the fire, and a way to transfer the heat to where it is needed, i.e., the cooking pot, griddle, or oven. Well designed, a rocket stove can cut fuel use by half and eliminate sooty smoke. Rocket stoves are not gasifiers, however, and consequently they can consume, but not produce, biochar.

Nathaniel Mulcahy, an award-winning industrial designer with a Ph.D. in fluid thermodynamics, used to work for large companies to develop home appliances. Eight years ago, he launched a small company in Italy called WorldStove® and set out to make a fuel-efficient, low-emissions, biochar-producing stove that could meet the cooking needs of villagers who lacked reliable fuel.

The best fuel for biochar at production scale is a local waste product that otherwise might be disposed of by burning in large open fires

Tribute to Lucia

WorldStove's Lucia model, their most popular home-kitchen stove, was named for Nathaniel Mulcahy's dog, who came to his master's side after a fall down a flight of stairs knocked him unconscious and shattered three vertebrae. If his head had flopped towards his shoulder, he would have died.

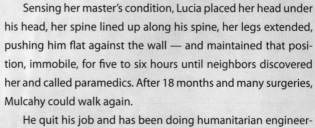

Sensing her master's condition, Lucia placed her head under his head, her spine lined up along his spine, her legs extended, pushing him flat against the wall — and maintained that position, immobile, for five to six hours until neighbors discovered her and called paramedics. After 18 months and many surgeries, Mulcahy could walk again.

He quit his job and has been doing humanitarian engineering ever since. The LuciaStove is his dog's memorial, designed to help millions of people and the environment.

Fig. 23: *WorldStove's Lucia model, named for the dog who saved the inventor's life.*

(generating carbon dioxide and black soot) or left to decay in piles (generating methane and other greenhouse gases). Mulcahy gives the example of a community kitchen in West Africa:

> Before we arrived in this one small town in Burkina Faso, the women's shelter was paying 3 euros per day for fuelwood imported from Ghana because they had no wood. Next to the women's shelter there was a shea butter plant, so there were tons and tons of karité shells or shea nut shells lying around. In that case, as in most places, the amount of waste material available far exceeds the need for cooking fuel. Egypt alone has 20 million tons of rice straw that they burn each year, which is equivalent to one-fourth of the country's energy use.

In some locations, especially in the tropics, grasses and shrubs may be easier and faster to convert for fuel than trees. Vetiver, a non-invasive tall grass, can be burned either raw (just leaves and stalks) or as cigar-shaped bundles, pellets, or briquettes. In Malawi, a typical family needs

4.4 pounds (2 kilograms) of biomass per day for cooking purposes, or about 200 yards of hedgerow.

In a Madagascar field trial, one of WorldStove's pyrolyzing stoves cooked the same amount of food as a standard charcoal grill, but required only 3 minutes to reach operating temperature, as contrasted with the charcoal grill's 20 minutes. The WorldStove used 1/20th of the fuel used by the charcoal grill, and generated 60 grams of 80-percent-carbon biochar.

The key to the WorldStove designs is their top and bottom plates, which Mulcahy builds to exacting specifications. The metal plates extract the gases from the fuel part of the chamber and generate Bernoulli-principle-driven venturis to create a negative pressure, while a flame cap based on Fibonacci spiral geometry prevents oxygen from entering the pyrolysis chamber. The design forces a clean, complete burn, with maximum carbon retention, and also produces a nitrogen-gas-charged biochar (the stove excludes oxygen but not nitrogen) that has a nearly neutral pH (7–7.5), making it ideal for agricultural uses.

WorldStove does not sell stoves in Africa. Instead, it helps local communities set up their own stove companies. Because of the precision in plate construction, required for fine-tuning the gas-air mixture, the plates are not easily crafted by rural village tinsmiths or masons, so WorldStove sends its plates to African partners, who build the stoves. With about 50,000 stoves projected for 2010, Mulcahy expects to be making 1 million stoves per year by 2020.

Fig. 24:

The LuciaStove life cycle.

In manufacturing "hubs" in rural villages, stoves are customized for local cooking traditions. Local production generates local jobs. By giving away all the potential carbon-credit earnings and allowing profit from the sales to remain in the community, WorldStove does not need to provide outside financing; the hubs can scale up quickly on their own marketing skills and earnings. WorldStove hubs are firing up now in Burkina Faso, Rwanda, Democratic Republic of Congo, Ghana, Togo, Senegal, Afghanistan, Uganda, Haiti, Colombia, Ecuador, and Mongolia.

A family that uses one of these stoves for cooking produces 300 grams of biochar per day, on average. That's 109.5 kilograms, or a tenth of a ton, per stove per year. Since about 80 percent of that char will be retained in the soil more or less permanently after being charged with compost, a quarter-ton of CO_2 is measurably sequestered per stove per year.[3] For every million stoves, villagers can sequester 250,536

WorldStove Five-Step Hub Plan

Step 1. Local group wanting the Hub provides building and personnel; WorldStove provides 3 large and 30 small stoves plus a small briquette press, free of charge.

Step 2. Once the Hub has demonstrated availability of all materials necessary to complete construction of 500 stoves, WorldStove sends a shipment of the first 500 critical components, necessary tools, and a small pellet press.

Step 3. Before a large press (600 kilograms per hour) is provided, the Hub must demonstrate orders for stoves or fuel.

Step 4. If they demonstrate that they are measuring, evaluating, and storing char, then the Hub can enter the carbon-credit program.

Step 5: Once the Hub has collected 5 tons of char, WorldStove will help the Hub develop afforestation and soil restoration programs.

Fig. 25. *The gas produced by a LuciaStove creates Fibonacci sequence spirals in combustion, similar to the patterns seen in the fruitlets of pineapple and artichoke, an uncurling fern, a pinecone, or the center of a daisy or sunflower.*

tons of CO_2 annually. Moreover, if one adds in the wood not being burned in open fires, since gasifying stoves are at least twice as efficient, WorldStove's million stoves are projected to supplant tree-to-CO_2 conversion by 38 million tons per year by 2020.

Mulcahy says:

Each time a biochar stove is used to produce a meal for a family of five, it can produce sufficient charcoal to filter 10 liters of water. This dramatically reduces the fuel needed to provide safe drinking water. Burying biochar, even after it has been used to filter water, helps restore soil nutrients and carbon, and it works especially well in Africa, which has the greatest capacity for carbon uptake. Furthermore, because the pyrolytic zone in a LuciaStove is the open air-fuel container, the stove uses the char it produces as a filter during the combustion of the gases produced, providing cleaner indoor air.

Converting manufacturing costs for the LuciaStove (about $12 per unit in Africa) into carbon-dollar terms, it costs $2.59 per ton to sequester carbon from biochar-producing cooking stoves.

Manufactured and distributed this way, wood-burning stoves, now an environmental problem, may become one of the world's greatest environmental hopes. Ramped up to full scale, just the reduction of inhaled soot would be the equivalent of eradicating malaria from the world.

Fig. 26: *By attaching a Stirling, Rankin, or Minto-cycle engine, electricity can be generated from the heat of a pyrolyzing stove.*

	 Aprovecho StoveTec GreenFire Rocket	 Top-Lit Downdraft Stove	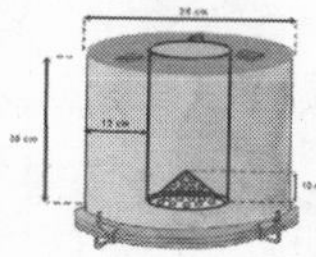 Anila or Retort Stove	 Lucia WorldStove
Adding Fuel	Can add fuel while operating.	Cannot add fuel while operating.	Cannot add fuel while operating.	Can add fuel while operating.
End Product	Mostly ash, with some char and torrified biomass. Not designed for biochar.	Mostly char, with some ash and torrified biomass. Tends to have higher pH.	Mostly char, with some ash and torrified biomass.	In pyrolytic mode: char; in gasifying mode: ash. The pH is adjustable.
Cook Time	The stove can be continually operated.	Depends on primary air supply settings, dimensions of the model, fuel load, and type of fuel used. Some models can only cook for 15 minutes.	Must cool for hours before refueling.	The stove can be continually operated until the char level fills the combustion chamber.
Stove Life	Not reported.	Not reported.	More than 10 years for burn enclosure, shorter for pyrolysis chamber.	Guaranteed 5 to 10 years.
Fuels Used	Woody biomass between ¼ inch and 5 inches in diameter.	Any small dry biomass (e.g., rice husks, small bits of wood, animal droppings).	Any small dry biomass. In the beginning, emissions are comparable to a standard stove; 30 to 50% of fuel used is burnt to ash.	Any small dry biomass; uses fuel considered too small to be used in other stoves.
Stove Maintenance	Easy to clean.	Information not available.	Separate burning and pyrolyzing chambers have different requirements.	Can be disassembled to clean and replace consumable components.
Removal of Residual Product from Stove	Can be removed with trowel, or stove can be tipped when cool.	Most commonly done by inverting the stove. Some models have handles to invert the stove while it is still hot.	Accomplished by inverting the stove. Some models have handles to invert the stove while it is still hot. Char must be	Most commonly done by inverting the stove. Some models have handles to invert the stove, or a slide drawer.

Fig. 27: *Comparing stoves.*

	Aprovecho StoveTec GreenFire Rocket	Top-Lit Downdraft Stove	Anila or Retort Stove	Lucia WorldStove
Removal of Residual Product from Stove		Char must be quenched or placed in a separate container.	quenched or placed in a separate container.	
Safety	"Safer to operate than the three-stone fire." Metal around door is sharp; using feed grate extends obstruction radius.	Some models have built-in tripods to increase stability.	Weight and center mass are designed to reduce risks.	A low center of gravity increases stability.
Flame Control	Adjustable damper below fuel intake.	Some models have adjustable primary and secondary air openings. Some have fans.	None.	Some models have adjustable valves to regulate heat output and fuel consumption. Some have fans.
Shipping Costs	Shipped for easy assembly from Ningbo, China, to any port in the world, in shipping containers with 1500 to 3000 stoves.	Shipped assembled, or assembled onsite.	Shipped assembled.	15–18 stoves are packed flat in a container the size of an assembled stove.
Manufacturing	At factory in China.	Usually manually built and assembled near the point of use.	Designed for mass production; requires aluminum casting.	Most models have been designed for mass production, which can be done locally.

BOOK IV

GARDENING THE EARTH

The absolutely essential proviso is that fossil fuel use is concurrently, progressively reduced to negligible quantities. Otherwise our newfound climatic stability from enhanced soil fertility will be short-lived and then total destabilization will be utterly irreversible. We can only play the soil fertility card once.

— Alan Yeomans

No solution is a solution for very long if it does not take into account the larger context of the problem it addresses. For biochar, or any human intervention into climate dynamics, to offer hope of saving the human experiment from catastrophic tipping points, we must stand back and look at the big picture. Biochar is a stabilization "wedge" that can be employed in coordination with other wedges to correct the skew in our carbon balance sheet. It can't do it alone, and it can't do it haphazardly.

Entropy being tirelessly at work to disperse energy and matter, concentrations of virtually anything have, since the dawn of history, been a source of wealth. Concentrations of minerals, heat and hydrocarbons confer wealth on contemporary states just as concentrations of fertile soils, forest spices, fish and game animals created wealth in earlier eras.

The genius of the makers of terra preta was that they concentrated nutrient flows from rivers, lakes, estuaries, fields and forests into the places where they prepared food, lived, and tilled the soil. Instead of dispersing those nutrients away by burning, commerce or erosion-prone tillage or water disposal, they conserved the elements that are needed to confer soil fertility.

In the modern world, by contrast, soil elements are dispersed from where they originate. Commerce is the principal export mechanism and it has become a monkey on our backs. Debt has become an index of wealth, and now, in 2010, we are required to calculate it in quadrillions instead of mere trillions, because exponential growth requires ever-higher debt. However, our virtual economies are liens on natural resources and the real physical assets to meet our demands are either being quickly depleted or no longer exist. That there is not enough Earth to satisfy these quadrillion-scale debts seems to elude most economists, but sooner or later we will discover the boundaries established by a physical universe

Carbon can be only stored in one of four pockets: earth, air, water and life. To keep our planet habitably cool, we need to store it mostly in the earth, its largest and safest repository. Our global economy has a way of taking it from earth and moving it to one or more of the other three pockets. Air is now overflowing with carbon, and is responding by overheating. Oceans are over-saturated and are acidifying. Life — animals, vegetables, insects and all the rest of it — is a very small and leaky pocket, and is becoming much smaller and leakier by the day, but only by way of that life reservoir can we stuff carbon into the deeper earth reservoir in a cost-effective way. We'll do that by investing more carbon into growing plants, over more of the surface of the planet.

24

Milpas

T HIRTY YEARS AGO, an ethnographic researcher named Ronald Nigh wandered up into the Mayan Petén region of Southern Mexico and began taking a hard look at the Lakandon Mayas' milpa system. Nigh discovered, on close analysis, that what European and Western anthropologists, foresters, and agronomists had been deriding for centuries as primitive "slash-and-burn" agriculture was actually a very sophisticated multicropping agroecosystem that for millennia had provided the Lakandon a rich, diverse diet while preserving and enhancing the fragile soils and tropical biodiversity, and providing a host of other ecological services.[1]

"Here's where we have to be thinking deeply," Kansas farmer Wes Jackson says. "Agriculture had its beginning 10,000 years ago. What were the ecosystems like 10,000 years ago, after the retreat of the ice? Those ecosystems featured material recycling, and they ran on contemporary sunlight. Humans have yet to build societies like that. Is it possible that embedded in nature's economy are suggestions for a human economy in which conservation is a consequence of production?"[2]

Jackson was mistaken. Humans *have* built societies in which conservation is a consequence of production. It seems that while the Sumerians, Babylonians, Assyrians, and Macedonians were busy salting the Fertile Crescent and the herders of the Sahara and Mongolian Steppes were overgrazing their way into the maw of some pretty nasty famine cycles, indigenous peoples of the Americas were quietly striking a deal with Mother Nature.

Ronald Nigh discovered such a place when he went into the Petén to study the Lakandon milpas.

Agroforestry combines trees and shrubs with annual crops and live-stock in ways that amplify and integrate the yields and benefits beyond what each component offers separately. Like other methods of sustainable agriculture, it is based on observing productive natural ecosystems and mimicking the processes and relationships that make them resilient and regenerative.

The Lakandon have been practicing such a system since they arrived in their part of the Petén following the fall of the final Mayan empire to the Spanish in 1697. Linguistic evidence and mitochondrial DNA show the Lakandon Maya to be of the Yucatecan Maya lineage of the Ah Itzá. Historically, the Itzá were a fierce people who remained in the Yucatán Peninsula long after other Mayan dynasties had collapsed. In the Classic Period, they erected the city of Tikal near Lake Petén Itzá in Guatemala, then abandoned it to migrate north. From their new capital at Chichén Itzá, which resembled Tikal in its monumental architecture arranged as an astronomical calendar, the Itzá established a new ocean-going trade empire reaching to the Aztec capital in the north and as far south as Honduras and possibly South America.

Eventually the Itzá lost a power struggle between three Yucatecan lineages all claiming to have descended from the Toltecs, and around 1331 they abandoned Chichén Itzá and returned south to the Petén, today part of Guatemala and Mexico near the western border of Belize. The Lakandon Maya speak Yukatek, a language brought south by the Itzá. There are only around 800 of them, in three villages in eastern Chiapas, Mexico. They are the smallest but oldest Maya group, sharing their original Itzá territory with some 500,000 Tzeltal, Ch'ol, and Tojolabal Maya who have migrated in from nearby highland areas.

Nigh found the sites where the Lakandon Maya dispersed their dwellings and gardens between the long, low ridges beside the Lacantu´n River, in undulating forested land. Abundant small streams provide fish and fresh water, sustained by nearly 100 inches of rainfall per year, and the climate is tolerably warm with little seasonal variation — ideal growing conditions.

The Lakandon practice an ancient system of perennial polycultures mixed with the production of maize, beans, squash, peppers, and some other annuals, according to taste and the local ecology. At the heart of the system is the milpa, a swidden (burned clearing in woodlands) in a rotation of stages. There are really two types of milpa practiced: "traditional" and "conventional." Traditional milpa is highly diverse, intensively managed, and high-performance, and was probably developed by the Itzá and other Maya groups of the past. The conventional form is a modern adaptation, with briefer recovery periods and more emphasis on corn production.

Earlier anthropologists and agronomists dismissed the Mayan milpas because they saw only the short growing periods for corn, followed by long fallow periods. There was a fresh clearing created by burning and then another short period of exploitive farming until the canopy closed and the site was abandoned. The actual practice, Nigh realized, was almost exactly the opposite. Not just the Lakandon, but all the Maya groups in the Petén share a deep knowledge of the forest. Each group has its own history and style, but all employ complexed agroforestry systems.

Prior to the arrival of steel axes and saws from European traders, cutting tropical hardwoods involved such an enormous expenditure of labor that it was completely impractical except where particular woods might be needed, such as cedars for canoes. Even in those cases, the favored alternative to hacking away for weeks with a stone blade would have been to burn the large tree at its base, fell it, and hollow it out with further burning and scraping.[3]

So, naturally, when a cornfield was needed and a windfall of large older trees on a level piece of ground was not forthcoming, the Maya burned a clearing. This had three benefits. First, as we have seen, it produced charcoal, some of it at the right temperature to make biochar, which would have gradually accumulated in soil layers, been digested by earthworms, and been transformed into terra preta. Secondly, it cleared the canopy enough to allow sunlight to penetrate to the ground, permitting annual food crops to be planted for a few seasons. Thirdly, it allowed the farmer to select cultivars of the plants most desired for succession after the annuals phase was over. These might be protected from the fire if they were low shrubs; the farmer could also gather and re-disperse favored seeds or encourage remnant trees with pruning and weeding.

Nigh observed that under Lakandon management, canopy closure is achieved in two to three years, rather than up to 10 as described by ecologists, through the propagation of fast-growing pioneer trees. Bats and birds are attracted to these pioneer species and bring the seeds of more shade-tolerant trees. Although most Lakandon farmers would probably prefer to prolong the forest-growth stage for a few more years, the site is ready to be reconverted to milpa at around 12 to 15 years.

Why would the Lakandon farmers prefer to prolong the forest-growth stage? It's the most productive part of the cycle when it comes to food, fiber, and animal products. The corn and beans are useful to exchange for ready cash, but compared to the yields that come from mature trees, those market crops are a lot more work.

Traditional milpa practices use techniques that add to soil fertility. Small fires of weed piles build up char in the soil and prevent hotter forest fires that would leave only ash. Besides planting legumes, which contribute nitrogen and beneficial nematodes, the Lakandon are fond of the balsa tree, which drops rich leaf clutter favored by soil micro-organisms and whose star-shaped, aromatic blossoms attract birds, bees, and bats. These pollinators transfer the seeds of food trees such as ramón (*Brosimum alicastrum*), hog plum (*Spondias mombin*), and Santa Maria (*Calophyllum brasiliense*).

Christopher Nesbitt has been growing food crops in the ancient style at his Mayan Mountain Research Farm near the town of San Pedro Columbia, in Southern Belize, for the past 20 years. He mixes some fast-growing native tree species, some annual crops, and some intermediate and long-term tree crops, to build soil and produce continuous harvests. Some of his trees are leguminous and hold nitrogen by the microbial attraction of their roots. Some are pollinator-friendly and attract bees and hummingbirds to transfer the fertile pollen of important food plants. Understory trees such as coffee, cacao, cassava, allspice, noni, ginger, and papaya benefit from intercropping with high-canopy trees such as breadfruit, açaí and coconut palm, cashew, and mango.

Fast-yielding crops like avocado, citrus, banana, bamboo, yams, vanilla, and climbing squashes provide an income for the farm while waiting for the slower harvest of samwood, cedar, teak, chestnut, and mahogany to mature.[4]

The World Agroforestry Centre reports that methods such as these can double or triple food yields per acre while reducing the need for commercial fertilizers. If best management practices were widely used, by 2030 up to 6 gigatons of CO_2e could be sequestered each year using agroforestry, which equals the current emissions from global agriculture as a whole.[5]

Chinampas

MEXICO CITY, CURRENTLY THE THIRD LARGEST CITY IN THE WORLD by population and eighth wealthiest city by Gross Domestic Product (GDP), is built on and around the ancient Aztec capital of Tenochtitlán. The settlement was originally fashioned on a small island in Lake Texcoco in the 14th century by the Mexica, a Nahuatl-speaking people who migrated into the high valley after the fall of the Toltec Empire.

Beginning in 1428, the Aztec Triple Alliance of Tenochtitlán, Texcoco, and Tlacopan remodeled the island and its surrounding hamlets into something much different. When the conquistadors first viewed Tenochtitlán in 1521, they were dumbstruck. Nothing in Europe rivaled the Alliance capital for the size of its monumental architecture, wide boulevards, and grandeur. Hernán Cortés marched across the gleaming white causeway leading into the city, and the Emperor Hueyi Tlatoani Moctezuma II came out from his palace to greet him and exchange gifts.

It did not end happily. Cortés kidnapped and ransomed Moctezuma. After a two-year struggle and siege, smallpox claimed the city, and Moctezuma's grand palace was razed by the Spanish to be replaced by Catholic churches and souvenir shops.

Mexico City has a very curious topography. Mountains and volcanoes that reach heights of over 16,000 feet (5000 meters) rise from a

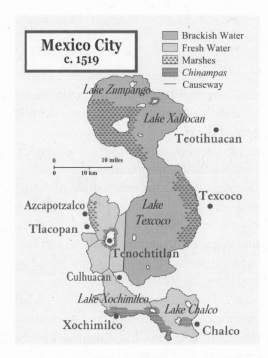

Mexico City
c. 1519

Brackish Water
Fresh Water
Marshes
Chinampas
— Causeway

Lake Zumpango

Lake Xaltocan

Teotihuacan

0   10 miles
0   10 km

Azcapotzalco

Tlacopan

Lake Texcoco

Texcoco

Tenochtitlan

Culhuacan

Lake Xochimilco

Lake Chalco

Xochimilco

Chalco

Fig. 28: *Mexico City Chinampas, c. 1519.*

6500-foot-high valley floor (higher than the Appalachian Mountains in North America). The valley is a bowl with no natural drainage outlet for the waters that flow from the mountains.

The Aztec Triple Alliance built dikes to separate incoming mountain water from brackish lake water. They used the fresh water to supply their neighborhoods and raise crops. Around the central city, and in the freshest parts of the lake, they developed an aquaculture system known as "chinampas," which harvested the mineral-rich silt deposited at the lake bottom and used it to make artificial islands where half to two-thirds of the food and medicine for the 200,000 residents was grown. *Chinampa* comes from the Nahuatl word *chinamitl*, meaning "square made of canes." Various grasses and bulrush that grew in the shallows beside the chinampas were used for baskets, mats, and clothing.

A typical chinampa was a 100-foot by 8- to 10-foot strip, separated from the next island by a canal about twice that width. These artificial islands produced three crops each year and were continuously replenished by dredging crews using roped buckets to scoop up nutrient-rich mud from the bottom of the canals, and carry human waste from the city to deposit on the islands. Huxley's Lord Edward Tantamount would be most pleased.

Besides bringing their food supply closer to the urban center, the chinampas agriscape provided drought resistance through sub-irrigation to the roots of the plants by capillary action. Willows, cherry, hawthorns, and other thirsty trees were planted at the margins of the islands to prevent waterlogging of the soil.

Similarly engineered food islands, some up to 60 feet (18 meters) in height and up to a mile wide, complete with ant-defeating ring trenches, were recently discovered in Amazonian floodplains of Bolivia, dating back to 1000 BCE. They cover an area of 200 square miles and were in continuous use until European contact. The first domestic peanuts, broad beans, chili peppers, rubber, tobacco, cacao, peach palm, and cassava may have been cultivated on these artificial islands interlinked by long, straight, elevated causeways.

For a high-altitude city like Mexico, chinampa gardening offered still another advantage: it ameliorated the effects of cold winter and hot summer temperatures. The giant lake served as a thermal battery, moderating and frost-proofing the climate on the islands.[1]

In the 16th century, city engineers opened canals and tunnels and drained the lake to prevent periodic flooding. The entire lake bed is now paved over by Mexico City. Today, without the lake and with only remnant chinampas and canals near Xochimilco, temperatures in the city can climb to 32 °C in summer (90 °F) or precipitously drop below freezing during half the year. The third largest city in the world has become vulnerable to heat and cold waves, drought, and famine.

Worse, the million-year-old lake will not be denied. It chugs the sewers of Mexico, chases with stormwater from the streets, and sits down to carve a new lake from the city streets, alleys, and vacant lots.

In 2006, agriculture accounted for only 3.9 percent of Mexico's GDP, down from 7 percent in 1980, and 25 percent in 1970.[2] While Mexico leads the world in production of avocados, limes, and sunflower seeds, it has become a net importer of beans, rice, soybeans, wheat, and even maize. With toxic pools of water, no chinampas, and a population now a hundred times larger than Tenochtitlán's, Mexico City subsists on a three-day supply of food for 19 million people.

You can't eat GDP.

Trees

WHEN EUROPEAN DISEASES SWEPT ACROSS THE NEW WORLD, forests replaced the cities, villages, fields, and roadways. Trees and vines grew impenetrably thick in the rich soils left by millennia of ancient carbon farmers. Forests emerged so rapidly they lowered global temperatures for hundreds of years.

Healthy forest ecosystems constantly absorb carbon dioxide from the atmosphere and store it in tree trunks, leaves, roots, soil, and woody debris. They sequester carbon for considerably longer periods than the average life cycle of a tree. Even a dead tree can continue to sequester carbon if it becomes part of a building that endures, is charred by fire, or is buried in a preserving environment.

With over 100,000 species, trees could be a quarter of all living plants. They give us ice cream, chocolate, paints, batteries, clothes, and cement. And yet, like Easter Islanders who cut down all their trees unthinkingly brought about their own extinction, we are still leveling our forests. In the tropics, 50,000 square miles of forests are destroyed each year by land-use changes. Equatorial warming of four degrees (coming as early as 2050 by some estimates) may remove these rainforests entirely, replacing them with grass savanna.

If we wanted to plant regionally specific trees everywhere there is suitable land for them, and protect them until they create their own local climate and favorable soil biology, what would it take? If many such efforts were undertaken, all over the world, what would be the effect on the carbon cycle? Fortunately, there are people and institutions perform-ing this work, monitoring its progress, and making those calculations.

South of The Gambia, in Casamance, Senegal, Nicolas Métro created the "Trees and Life" project to help the Jola people replant mixed-use (fruit, timber, fuel, habitat) indigenous trees around villages where deforestation, partly the result of a 16-year civil war, was most severe. In Casamance, 20 percent of the people do not have access to electricity, 40 percent lack running water, and 84 percent of the chil-dren under age five are malnourished.

The framework Métro devised had several phases. First, the villagers, principally women, were brought into constructive dialogue, which he calls "participative afforestation." During this phase, the value of trees was discussed and decisions reached, with expert advice, about which types of trees should be planted and how best to protect them. Trees and Life, in cooperation with the local government, provided seedlings and materials for a village tree nursery, staffed by volunteers.

Secondly, the villages were assisted in construction of rainwater col-lection and retention systems to provide water for the trees during the dry periods, as well as for other village uses. Sheaths were provided to protect the transplants from animals and wind, and expert teach-ers came from France to provide technical assistance in planting the trees and assuring their survival for the first year. Between July and September 2009 (the Senegalese winter), 300,000 trees were planted. That may seem like a large number of trees, but the goal was far more ambitious — 15 billion trees over 10 years. The training in 2009 involved 130,000 people in 90 villages. That works out to only slightly more than two trees per planter.

The real seeds were being planted in people. About one-quarter of the world's forests are managed by 200 million community members and families. Forty percent of the forest in the Northern Hemisphere is owned by 30 million families. Twenty-five percent of the forest in the South is owned or managed by families or communities.

Forests can be managed sustainably or unsustainably. Until now, the second option has all too often prevailed. As James Lovelock likes to say, "A tree plantation is not the same as an ecosystem." Forest ecosys-tems provide food, medicines, diverse animal, plant, and insect habitat, air conditioning, soil retention, energy, recreation, clean water, and many other services. Tree plantations provide only short-term cash and a timber product.

Some of the species of life that share this planet really want to be our friends, and have gone to great evolutionary lengths to prove it. The leaves of *Moringa oleifera*, the horseradish tree, originated in the southern foothills of the Himalayas in India, near the source of the sacred Ganges. By the end of the Sumerian Empire, it had spread to Pakistan, Bangladesh, Sri Lanka, and Afghanistan. Today, it is widely cultivated across Africa and Central and South America, Malaysia, Indonesia, and the Philippines. Moringa grows best in dry sandy soils and is especially well-suited for drought areas, since the tree is in full leaf at the end of the dry season when other foods are typically scarce. Propagated either from seed or by planting limb cuttings, the tree starts bearing edible pods in six to eight months.

It has more Vitamin A and beta-carotene than carrots, more calcium than milk, more iron than spinach, more Vitamin C than oranges, and more potassium than bananas.

The many uses for just this one species include alley cropping (biomass production), animal forage (leaves and treated seed cake), biogas (from leaves), domestic cleaning agents (crushed leaves), blue dyes (wood), fencing (living trees), fertilizer (seed cake), foliar nutrient (juice expressed from the leaves), green manure (from leaves), gum (from tree trunks), clarifier of honey and sugar cane juice (powdered seeds), honey (flower nectar), medicine (all plant parts), ornamental plantings, biopesticide (soil incorporation of leaves to prevent seedling damping-off), pulp (wood), rope (bark), tannin for tanning hides (bark and gum), and water purification (powdered seeds and charcoal).

Moringa seed oil (yield 30–40 percent by weight), also known as Ben oil, is a sweet, non-sticking, non-drying oil that resists rancidity. It has been used in salads, for perfume and hair-care products, and as a sewing machine lubricant. The high protein seeds are eaten green, roasted, powdered, curried, or steeped for tea. Leaves, flowers, seeds, pods, roots, bark, gum, and oil have medicinal properties and have been used since the times of the Egyptians and Greeks.[1]

Because African villagers are familiar with it, moringa will likely be among the quarter-million trees that Trees and Life will start in each of the 90 villages in Casamance over the next three years. Eventually Trees and Life will create six central tree nurseries to support the network of 60 village-managed nurseries. From each women-directed village cooperative, the organization will offer production and marketing support and microfinance for local forestry enterprises (fruit products, beekeeping, gum arabic, medicinal extracts, salves and decoctions, guinea fowl, etc.).

For Trees and Life, the final phase will be monitoring and reporting the progress and integrating the data into the Total Cooling Project of the NASA Land Atmosphere and Resilience Initiative. As added incentive, a prize is being offered to the village whose survival rate for the trees is strongest after three and five years, respectively. Métro told a gathering at the Climate Summit in Copenhagen that he hopes to determine the impact on temperature and hydrological cycle at the local, regional, and global level. His home organization in France, Kinomé (Japanese for "the viewpoint of the tree"), is already planning the next greater scale for Trees and Life in the tropics; 15 billion trees is just a starting point. At the conference in Copenhagen the government of Senegal pledged $2 million to build a "Great Green Wall" of drought-adapted trees through 11 countries south of the Sahara, crossing the continent from east to west. The wall would be 9 miles (15 km) wide and 4,831 miles (7,775km) long.

The Kinomé plan is that eventually the villagers can be paid for the carbon sequestered by the trees and for the "carbon avoided" by protection of the existing forests that would have been deforested. Ground readings taken in several forests near Casamance recorded 260 tons of CO_2 per hectare, including 74 tons in aboveground biomass (30 percent) and 186 tons in the soil (70 percent). That survey provides a useful baseline from which to calculate family-managed carbon sequestration services in the future and to reward real success.

In 2004, the Nobel Peace Prize was awarded to Wangari Maathai, a Kenyan environmental and political activist. In the early 1970s, it became evident to Maathai that the root of most of Kenya's social problems was environmental degradation. She connected her ideas of environmental restoration to providing jobs for the unemployed by founding Envirocare, a company that employed people to plant trees. On June 5, 1977, marking World Environment Day, she led a march from downtown Nairobi to the outskirts of the city where she planted seven trees in honor of historical community leaders. This was the first "green belt" in what would become the Green Belt Movement. Maathai encouraged Kenyan women to start tree nurseries using heirloom native species. With money from the United Nations Development Fund for Women, she was able to pay a small stipend to the women who planted seedlings and to the husbands and sons who were literate and able to keep accurate records of success rates.

Wangari Maathai's struggle — in which she was repeatedly arrested while planting trees, beaten by police and paramilitary groups, placed under house arrest, besieged in her home, and periodically forced into exile — was ultimately vindicated when, in 2002, her Rainbow Coalition defeated the ruling party of Kenya. Within her district, Maathai won 98 percent of the vote. She became Assistant Minister for Environment and Natural Resources and founded the Mazingira Green Party of Kenya to support candidates who embodied the ideals of the Green Belt Movement.

In the 1990s, when the Green Belt movement planted 20 million trees, that seemed like a very large number. In 2006, the UN Environment Programme launched a "billion tree campaign" for world plantings by the end of 2007. That goal was surpassed in 2005 and had to be raised to 7 billion by the end of 2009. Plantings of 6.3 billion trees were achieved, but no one knows how many of those trees actually survived. In Palestine, where Murad Al Khufash led an effort to plant olive trees under the sponsorship of a trees-for-air-miles program, two years of plantings were chain-sawed to make room for Israeli settlements in the West Bank. Such experiences are not uncommon in places where the value of forests is still of far lower priority than population and commercial demands.

Still, tree planting has been gaining momentum, and a friendly rivalry has begun. Pakistan has apparently set a record for aquatic tree plantings, with volunteers wading through chest-high water and knee-deep mud to plant an average of 1800 mangroves per day, each. In a single day and without any mechanical equipment, volunteers planted 541,176 young mangroves in the Indus River Delta to best the previous Guinness World Record of 447,874 trees in a day, held by India.

Richard Garstang, head of the program, said, "We hope that tree planting competitions will become as popular as cricket matches."

The Power of Youth

CHINESE YOUTH CLIMATE ACTION NETWORK is the largest youth organization in China devoted to climate change. Working to green universities, CYCAN set a goal of 20 percent reduction of greenhouse gas emissions from higher education institutions by 2012. One of the things it learned along the way was the value of trees.

In 2001, CYCAN joined the 21st Century Greater Beijing Reforestation Model project aimed at preventing desertification in China, then already encroaching on the outskirts of Beijing. By the end of 2009, program affiliates, local residents, and volunteers had planted trees on 7000 acres (2800 hectares). The program attracted the attention of Toyota Motor Corporation, which sponsored 100 Japanese students to join CYCAN China Youth in planting 450 Chinese red pine trees in the Hebei Province of China. Work during the day was followed by parties at night, where language barriers fell to the pulse of worldbeat. Another Japanese company, Toshiba, funded tree planting activities at 12 sites in Japan and six overseas sites, helping China Youth plant 600,000 trees.

Youth exchanges of this kind make tree planting sustainable by making it fun. It can be enormously effective as well. Consider this: According to the most recent US census, as of 2008 there are 16,715,000 high school students and 18,632,000 college students, or 35,347,000 students, in the US. Their school year is 180 days. Reserving a few holidays, they're left with 180 days of vacation. If every high school and college student in the US were to spend his or her vacation time planting trees, each could start 1000 bare-root seedlings per day.

However, it is not all just planting; each planter needs a supporting crew, roughly two nursery workers and one fertilizer-waterer-weeder to ensure good seedling growth and low mortality rates. It would also be nice to put a little fungal-enriched biochar in the soil below every rootstock. Of that 35-million-person workforce, 8.8 million might be designated as planters, and the others could be support crew, although certainly rotation between the various tasks would keep it more interesting. At 1000 bare root seedlings per planter, the students could plant 8.8 billion trees per workday. That's 1.6 *trillion* trees per school year.

What would that do to the atmosphere? That is a tough question because there are many variables. To begin with, the photosynthetic efficiency of forests varies with species, ecological system, and climate. And, recalling what James Lovelock said about plantations not being ecosystems, we would want to be cautious when generalizing about numbers of trees and areas of land, because the greater goal would be to optimize for biodiversity rather than maximize for timber production.

What happens to that harvested carbon? If you make biochar, then 50 to 80 percent would be sequestered for about a thousand years. And the unharvested part? The green, aboveground biomass and the brown, belowground roots and mycorrhizae double the total carbon being held.

In one summer vacation of 180 days, US high school and college students could plant 630 million hectares (1.6 billion acres). If you remove the water-covered areas of the contiguous 48 states from consideration, there remain 7.66 million square kilometers (19.15 million square miles) of land surface. It would take those students only 7.24 months to plant that with new trees. Calculating roughly for the diverse temperate forests of North America, each hectare would store 289.2 tons of CO_2 annually.

I asked Frank Michael, a physicist studying forest growth, a hypothetical question. "Suppose everyone in the world planted one tree every day. How long would it take to restore the balance of carbon in the atmosphere?" He said:

If one person out of 2000 hand-planted 2000 trees a day, that would be the equivalent of everyone in the world planting one

Step-Harvest for High Planting Density

Frank Michael is a program director at Global Village Institute, studying forest growth dynamics. He previously worked at United Aircraft Research Laboratories, doing chemical analysis of gases using infrared and optical spectrophotometry and helping to advance laser holography, interferometry, and coherent optics. Before that he was at NASA's Ames Research Center, investigating simulated spaceship entry into the atmospheres of other planets, and doing instrumentation for the flying solar observatory. Now he crunches numbers on trees.

Michael has calculated that one person with a dibble stick or hoedad can plant 1000 to 3000 bare-root seedlings per day. This has been borne out by groups in the field such as Trees and Life and China Youth. For a high planting density using Michael's innovative step-harvest method, planters separate themselves two meters abreast (about 6 feet) and walk forward, each one planting seedlings. While topography varies, they plant generally in "unit cells" measuring 8 meters by 8 meters. Each hectare (2.47 acres) will have 156 cells, each planted with 16 seedlings — 2496 seedlings per hectare. (We are using metric units here because it makes the calculations easier).

The essence of the step-harvest method is successive harvesting of one-half of the evenly spaced remaining trees until only one tree is left. The process yields six times the carbon density of the forest and its products, compared to clearcut harvest at multi-decade intervals. Workers would thin out one-half of the new trees at six years of age to make space for the larger girth of the remaining trees. The culling would be the start of mixed-species, mixed-age management. Selection criteria would be based on health of the forest — leaving the healthiest and best-positioned specimens and removing the weak and crowded ones. When planted with the kinds of trees appropriate to the diverse biomes of North America (and accounting for changing climate), each of the 64-square-meter cells will take, on average, about 24 years to complete one step-harvesting cycle.

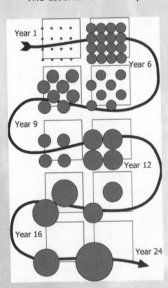

Fig. 29: *The step-harvest forestry method.*

tree per day. If we also employ a crew of two to grow seedlings, and another person to prepare the soil and to water and weed around the seedlings, then one person out of 500 would have to work on a crew planting 2000 trees a day in order to equal one tree per day for every person in the world.

What area can one planter (and associated crew) cover in one day? It depends on the density planted, which depends on the type of plantation. Assume that the average tree species reaches the shoulder of its biomass production curve in 24 years and has a canopy diameter of about 24 feet. Selected harvesting can begin after canopy closure of different-aged trees within the cell. The trees can be managed in one of two plantation modes.

Michael paused to write on the chalkboard:

Conventional: 1 tree/24' x 24',
yielding 24 mass units/24 yr = 1 mass unit/yr
Step-harvest: 16 trees/24' x 24',
yielding 148.5 mass units/24 yr = 6.2 mass units/yr

With the step-harvest method, he explained, one-half of the trees are successively harvested at their canopy closure, in four or five harvests at years 6, 9, 12, 16.5, and optionally in year 24. The area is planted with mixed species and evenly thinned, rather than planted with a big open space between trees in anticipation of their eventual size at maturity. The step-harvest method yields more than six times the volume of biomass, and the harvest can begin at year 6 after planting. In fact, this first harvest yields the greatest amount of biomass. In the end, you could choose whether the oldest tree is harvested and the whole cycle repeated, or one tree per cell is left to become old-growth forest.

Michael calculates that 3.8 million people could plant one year's worth of carbon emissions annually. In 2009, when we spoke, that would have been 4.5 GtC. By this same arithmetic, 7.6 million planters could plant twice that many trees per year, and begin to *reduce* atmospheric CO_2. Michael says:

If everyone in the world planted one tree each day, by the end of the 2nd year, the sequestered CO_2 would exceed the global emissions. Meanwhile, the trees you first planted are now older and bigger. So by year 3, the sequestered CO_2 is more than three times emissions. By year 4, it is five times emissions. By year 5, we sequester seven times the emissions. By the end of year 6, we are annually sequestering more than ten times the 2009 CO_2 world emissions.[1]

The principal obstacle is not lack of manpower, however; it's the availability of land. Twenty-three million planters could plant enough trees to offset global CO_2 emissions in two months, but they would use up all the fallow arable land in the world.

Where do we send the tree planters after all the unused arable land is planted? Deserts cover a third of the Earth's surface. Climate change is causing deserts to expand at an accelerating pace. Expanding deserts disrupt evaporation and rainfall patterns, desiccate forests, and grow steadily larger, changing regional climate. And yet, what is true about desertification affecting climate is also true about de-desertification. By greening barren lands, the hydrological cycle is restored, ecosystems are re-invigorated, and carbon is steadily removed from the atmosphere.

Greening the Desert

IN 1983, GEOFF LAWTON PAID A VISIT to the Tasmanian farm of Bill Mollison to take his famous 72-Hour Permaculture Design Course. Lawton stayed, and eventually worked as farm manager when Mollison was away teaching and speaking. An eager experimenter, Lawton plucked gems of wisdom from different styles of agriculture around the world and applied those to what became the Permaculture Research Institute of Australia.

From the lowlands of Mexico, Lawton learned about chinampas, and he constructed several on Mollison's farm. From P. A. Yeomans, Lawton studied the keyline method. From Elaine Ingham and Paul Stamets in the Pacific Northwest, he learned how to make the conditions favorable to bacteria and fungi. Using mulch most favorable to fungal growth, he discovered how to hold both limited soil nutrients and seasonal moisture. He learned that deep roots grow good nematodes and 15 to 20 feet of uncompacted subsoil creates well-rooted plants. He learned that good soils contain 50,000 protozoa per gram, and those protozoa have to eat 500 million bacteria per day, a process that releases 400 million molecules of available nitrogen. He practiced his crafts with compost teas, mulching, and successional planting until the Permaculture Research Institute was world-renowned. By 1998, Lawton was advising projects in 25 countries.

Lawton got his big break in 2001 when the government of Jordan approved a proposal for a test farm in one of the least hospitable places on Earth for farming. In the Kafrin area of the Jordan Valley, 10 kilometers from the Dead Sea, the nearly flat desert receives only two or three light rainfalls in winter. The fine-grained silt is salty, and even the wells in the area are too saline to be used for irrigation.

It was there that Lawton and his team of permaculturists set up a small, 5-hectare farm and began digging swales — 2-meter-wide mounds and shallow trenches that crossed the farm in wavy lines on contour. They planted leguminous forest trees in the mounds to fix nitrogen and make leaf fodder. Each tree was given a drip-node from an irrigation line coming from a water dam built to capture road runoff; the water behind the dam was stocked with tilapia and geese, which contributed organic fertilizers for the trees.[1]

In the moist trenches, Lawton's team planted olive, fig, guava, date palm, pomegranate, grape, citrus, carob, mulberry, cactus, and a wide range of vegetables. Between the swales, barley and alfalfa were planted as nitrogen-fixing legumes and as forages for farm animals. Tree and vegetable plantings were mulched with old newspapers and cotton rags, and animal manure was added before and after planting. Chickens, pigeons, turkey, geese, ducks, rabbits, sheep, and a dairy cow all contributed droppings. They were fed from the farm once there were enough trees and plants growing to harvest regularly without overtaxing the system.

Within the first year, the soil and well water began showing a marked decline in salinity, and the garden areas had significant increases in growth. Pests were minor and largely controlled by the farm animals. The combining of plants and animals brought about the integration of farm inputs and outputs into a managed ecosystem of continuous production, water conservation, and soil improvement. In less than a decade, a permacultural balance had been achieved, with lessening inputs and improving outputs.

Agriculture is no more than the sum of individual farms. If a farm is true to its essential nature, in the best sense of the word, it is conceived as a kind of individual entity in itself — a self-contained individuality. Future methods to begin re-greening desert peripheries in a spiral of increasing fertility, whether by US students on summer break or by China Youth, would likely be variations on this permaculture model.

Sahara Forest

WITHIN RECORDED HISTORY, the Sahara and many other great deserts were covered with forests of cedar and cyprus, savannas, and diverse biomes. The Sahara is the product more of over-exploitation and abuse at human hands than of the climate changes that came in 100,000-year cycles. Recently a small consortium of companies joined to pursue a vision of a re-greened Sahara.

The Norwegian philanthropic Bellona Foundation set out to show that solar-powered seawater desalination and soil restoration in arid climates is not only a practical solution from a technical standpoint, but it can also be commercially profitable. "We need to go green by black numbers," said Bellona's Frederic Hauge. One of the ways of doing that was by harvesting elements and compounds from the seawater, such as calcium carbonate, sodium chloride, magnesium chloride, and lithium.

Bellona put together a team of three engineers:

- Charlie Paton, an innovative engineer from San Jose, California, has won awards for his work in solar-powered desalination. His brain-child, the Seawater Greenhouse, creates a cool, humid growing environment by evaporating seawater from cardboard grills in the front of the greenhouse and condensing distilled water at the back. The water is used to grow microalgae, seaweed, shellfish, shrimps, and fish, whose wastes are converted to fertile soil. In southern California, this solves two urgent problems: spreading deserts and chronic water shortages. Production of shade-house and mariculture crops such as fish and abalone is also possible, and aquaponic seaweed and algae can be converted to biogas and biochar.

• Michael Pawlyn established Exploration Architecture in 2007 to move architectural engineering more into the realm of biomimicry, patterning from nature. Pawlyn is best known for his enclosed biomes designs for the Eden Project, an environmental center in Cornwall, England.

• Bill Watts is a specialist in sustainable architecture and renewable energy for the Carbon Trust in Northern Ireland. His company, Max Fordham Consulting Engineers, designs low-energy housing, offices, sports centers, museums, and government buildings.

Pawlyn and Watts previously had designed the buildings and transport systems to surround Paton's desalination and aquaponics modules in California. With Bellona's support, they inaugurated the Sahara Forest Project in 2009. The first prototype, a 1000-square-meter building in Oman, evaporates 5 cubic meters of water per day. Scaled up, Hauge says, that would equal 50 cubic meters per hectare. The evaporation also removes 800 kW of heat from the buildings. Since it is only used in the daytime, the building can be easily powered by photovoltaic cells, with no batteries.

In the next scale prototype that the team envisions, 50-megawatt Fresnel-concentrated solar thermal collector towers will turn steam

Fig. 30: *The Sahara Forest Project greenhouses are arranged as a long "hedge" snaking inland from the coast, using concentrated solar-thermal power to desalinate seawater and provide electricity. Behind the greenhouses, orchards are planted in shelterbelts, and dryland crops such as jatropha improve the soil while creating humid microclimates.*

turbines to provide energy. Hauge estimates that a 10,000-hectare (24,700-acre) area of greenhouses will evaporate over 275 million gallons of seawater per day. The design can produce 450 to 550 tons of high-value horticultural crops per hectare per year and sequester 8 tons of CO_2 in soil. A 50-hectare "farm" in the Sahara would produce 34,000 tons of vegetables, employ over 800 people, export 155,000 kilowatt-hours of electricity, and sequester more than 1500 tons of CO_2 each year, while leaving behind a tropical forest and a naturally-carbon-sequestering climatic, hydrological, and soil regime.[1]

The Groasis Waterboxx

Another designer who has been devoting a lot of time to the problem of planting trees in arid climates is a Dutch engineer, Pieter Hoff. Beginning with the way seeds grow in nature — passing through the stomachs of birds and being excreted with a protective sheath of moist and nutrient-rich bird manure — Hoff designed a small plastic cylinder that could serve as a sheath around rooted seedlings to protect the tender plants from winds, daytime heat, nighttime cold, animals, and dry soil until they could establish themselves and send their roots deeper.

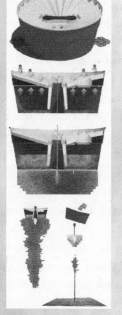

His 2003 invention, the Groasis waterboxx, draws water from air (without energy) and stores it during the day in an internal compartment. The box has a shaft up the center that surrounds, cools and protects the tree as it grows. A small wick allows moisture from the water compartment to be drawn through the soil to the seedling's roots. The top of the waterboxx is designed to fill the compartment during rain events and then slowly release its contents to the soil.

Once the tree is established, it no longer needs the waterboxx, so a worker collects the box, withdrawing the stakes that keep it affixed in high winds, discards the used wick, installs a new one, and the box is ready to return to service with a new seedling. A hemp or jute cloth is laid around the new tree in the place of the box in order to protect the soil from evaporation and to inhibit opportunistic weeds.

Fig. 31: *The Groasis waterboxx.*

Drey's Challenge

THE TRANSITION TO A CARBON-AWARE ECONOMY will lead the world away from fossil fuels and into greater reliance on renewables, including forest biomass. Sustainable forest management must become a key element in this transition if we are to avoid depleting yet another resource through a mad scramble for economic growth at any cost.

The good news is that sustainable forest ecosystem management may actually be more economically productive than the industrial tree farms and cattle ranches that supplanted wilderness old growth all over the world in the 20th century.

In 1899, the sawmill at Grandin, Missouri, east of the Current River, consumed 70 acres of woodland a day, and produced in excess of 250,000 board feet of lumber, lath, and shingles. Many sawmill towns came into existence in rural Missouri around the turn of the century, using river power, animals, or coal to turn the Ozark forests into timber and sawdust. By the 1920s, all but a few acres of Missouri's virgin forests were gone, the mills were shut down, and all those jobs were lost.

The erosion of the soil was so severe that farming also entered a crisis, and many farmers sold out and left. The streams were choked with silt, and wildlife declined to only about 2000 deer and a similar number of turkeys in the entire state. Eventually, in the 1930s, the federal government bought the barren landscape and sent in the Civilian Conservation Corps to fight fires, build roads, and plant trees.

Among those taking an interest in the Ozark forests was a young businessman named Leo Drey. In 1951, Drey used some of his inheritance from his father's business, Drey Perfect Mason jars, to purchase 1407 acres of oak and pine, much of it rotten, for $4 an acre. Drey was 34 years old and a 1939 graduate of Antioch College, whose founder, Horace Mann, had told students, "Be ashamed to die until you have won some victory for humanity."

Recognizing his own limitations, Drey hired a forestry instructor at the University of Missouri to help him locate and buy forested land. Over the next few years, he expanded his holdings to over 125,000 acres, including a single purchase in 1954 of nearly 90,000 acres from the National Distillers Products Corporation of New York. He called it Pioneer Forest, not only because he intended to manage it as a model of sustainable forestry, but also because Pioneer Cooperage Company of St. Louis had originally assembled the largest tract. Pioneer Cooperage had begun sustainable harvest of white oak for barrel staves under professional foresters Ed Woods and Charlie Kirk. The company was then sold, in 1948, to National Distillers, with its forestry team intact, but several years later the distilling company decided to liquidate the forest and sell out.

Woods and Kirk alerted Drey, who entered into negotiations to purchase the land and tried to save as much as he could, but National Distillers insisted on the right to cut all white oak over 15 inches in diameter. As Drey and his new forestry team watched helplessly, 12 million board feet of oak were cut in 1954 alone, and another 12 million were lost to the cataclysmic drought and fires of the era.

Much of the forest that Drey, Woods, and Kirk began with was severely degraded, but they maintained long-term records of inventory, species composition, stand volume, and other indicators of systemic health. Important for forest health and productivity, they believed, was to maintain diverse species and ages of trees on every acre. To do that, they practiced uneven-aged silviculture through single-tree selection, marking and taking out the weak, deformed, and crowded trees, and paying careful attention to slope, soil, and canopy openings to favor young reproduction and continued growth of the best trees of varied species and sizes. They were more concerned about what remained than they were about what was removed. They also kept logging equipment out of stream bottoms, maintained forested stream buffers, and rarely cut more than 40 percent of the volume in any given stand.

During the 1960s in the West and South and by the early 1970s in the Ozarks, professional foresters on federal and state forests and in forestry schools shifted to a more intensive commercial model of forestry based on even-aged management by clearcutting. Industrial-scale harvesting equipment often abused the terrain and watercourses. Regeneration or planting favored the most commercially valuable species, often mono-cultures of hybrids designed for fast growth. For decades, most forestry research, funded by the federal government or private industry, was directed to even-aged management to improve industry profitability. Until recently, little consideration was given to uneven-aged management or to ecosystem health, carbon sequestration, or biodiversity.

Many of the financial incentives that stripped the forests of Missouri in 1910 are still present in 2010. Landowners can get more quick cash by liquidating their forests than they can get by managing them.

The lasting accomplishment of Leo Drey and his small team of foresters was to prove that their uneven-aged management system was both sustainable and profitable. They proved they could secure natural reproduction of commercially valuable species while also producing a full array of ecological, social, and cultural values.

In 2002, the volume of the standing trees in Pioneer Forest was more than three times what it had been in 1952, and asset value had increased nine-fold since 1972. A 2002 study found that in the previous six years income had exceeded expenses by about 50 percent. For the landowner, the forest had almost certainly been as profitable as it would have been under an even-aged regime. Moreover, once an acre of Ozark forest is clearcut, it requires 80 to 100 years to regrow. During this same period, four or five harvests could be made under the uneven-aged system.

The difference between the two models is more than money, however. Every acre of Pioneer remains a true forest. In studies of salamander populations, microarthropods in leaf litter, black bears, and migrant songbirds, Pioneer Forest wins easily over even-aged forests, and the value for recreation is continual. The average turnover rate for Pioneer's canopy is 189 to 228 years. Its soils are a rich carbon sink, and it is more resistant to drought, disease, and pest damage than any monocrop plantation could ever be.

BOOK V

AT THE TURNING POINT

We are now at the stage when the Easter Islanders could still
have halted the senseless cutting and carving, could have gathered
the last trees' seeds to plant out of reach of the rats. We have the
tools and the means to share resources, clean up pollution, dispense
basic health care and birth control, and set economic limits in
line with natural ones. If we don't do these things now, while we
prosper, we will never be able to do them when times get hard.
Our fate will twist out of our hands. And this new century
will not grow very old before we enter an age of chaos and
collapse that will dwarf all the dark ages in our past.
Now is our last chance to get the future right.

— Ronald Wright

The Chinese symbol for "crisis" includes two concepts: danger and opportunity. In climate change, our primary danger is that we will cross some unseen,
irreversible threshold before we are aware of it or can muster a critical mass of

opinion to change our course. "Tipping points" are points in the functioning of a system (including the system of all life on Earth) where a small change can result in large effects. Changes in the system have little or no effect until a critical mass is reached. At that point, a further change can "tip" the system into a significantly different way of functioning.[1]

Few of the attempts to avoid the dangers of the climate crisis will matter if human behavior remains unchanged. Some in public policy circles have proclaimed, "Lifestyle is not on the table," meaning that governments and opinion shapers should not consider trying to change the way people behave. And yet, our human behavior is at the crux of our climate problem. What we choose to eat, how we work, how we use electricity, the buildings where we work and live, how we travel, and what we buy, all matter.

The writer Stephanie Mills observed:

Because the planet is finite and somewhat of a closed system, we all live intimately with the results of our acts. Things do add up and, as population grows, there are more of us adding to the adding of things. Similarly, the benefits from many individual actions of self-restraint, frugality, and material simplicity will add up.

We know from ocean sediments, ice cores, and other evidence that over hundreds of thousands of years the equilibrium between carbon dioxide input and removal has never been more than one to two percent out of balance, a strong indication of a natural feedback wisdom. Now, in just an instant of geological time — the industrial era of the past two centuries — that balance has been thrown out of kilter by large percentages and may now be approaching, or be at, an irreversible tipping point.

Human ingenuity was a leading cause of Earth's tip into imbalance. If it is not already too late, we humans may also hold the power of the tip back into balance. We need to come together, recognize the common threat, and act. There are many leverage points where, once the resolve to act is found, we can apply our collective abilities to get the kind of tipping point we want.

The Biochar Critique

THERE IS A GROWING INTEREST IN BIOCHAR, both in countries most threatened by the near-term effects of climate change and by many nongovernmental organizations that have labored in the fields of reforestation, regenerative agriculture, water conservation, and similar issues for years. There is also an opposing voice, and it is making itself heard.

In December 2009, the nations of the world gathered in Copenhagen, Denmark, to debate sustainable and socially just solutions to the climate crisis. The urgency had by then become obvious, and gone were many of the parsings of science and economics that had made for contentious UN gatherings in Bali, Poznan, and Kyoto. By 2009, the evidence had become all too clear that unless political rivalries were put aside, humanity as a whole faced a prospect as dire as that of the Maldives, the tiny island chain in the Indian Ocean that was slipping beneath rising seas. Indeed, the young president of the Maldives, H. E. Mohamed Nasheed, rose in Copenhagen to pledge his country would become carbon-neutral by 2020, and he called on every other country to do the same, "not just for sake of the Maldives, but for the sake of the entire planet."

How major industrial nations like the US, China, India, Japan, and Russia could become carbon-neutral, or even significantly reduce the growth of their carbon emissions, was the subject of considerable discussion. At earlier conferences, much hope had been pinned on the technological prowess of *homo sapiens* to invent its way out of the monkey trap without releasing the fruit in its paw. So, for instance, there had been discussions of carbon dioxide capture and sequestration at power plants and furnaces; fertilization of the oceans to stimulate phytoplankton blooms that would drop carbon to the ocean floor to become rock; and solar radiation management by means of reflectivity — bouncing the light of the Sun back to space with orbiting shades or mirrors, stratospheric sulfur aerosols, cloud reflectivity enhancement, and urban surface albedo (more commonly called "painting roofs white").

With each of these patches, engineers encountered the law of unexpected results. None of the techniques was without side effects, and most of those were bad news. Urgent studies were undertaken by the US National Academy of Sciences, NASA, the Institute of Mechanical Engineers, the Royal Society, and the British Parliament.

The Royal Society report in 2009 was a model in damning with faint praise. While advocating further research and standard setting, its opening page read, "The safest and most predictable method of moderating climate change is to take early and effective action to reduce emissions of greenhouse gases. No geoengineering method can provide an easy or readily acceptable alternative solution to the problem of climate change."[1]

The Society graphed each of the technical fix "wedges" on a chart that also included three new technologies — afforestation, carbon capture and storage at source (CCS), and bioenergy with carbon storage (BECS). The hands-down winner in terms of economy, feasibility, and safety, it turned out, was planting trees.

The new kid on the block, for the Royal Society, was BECS, which includes biochar. It says quite a bit about our fascination with science

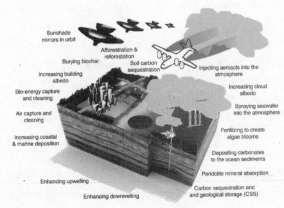

Fig. 32:
Geoengineering options include a range of choices for solar radiation management and removal of carbon from the atmosphere.

and engineering that an ancient style of farming could find itself lumped in with space mirrors and peridotite mines as a climate modification strategy, but the inclusion marks a milestone for sustainable agriculture. The Royal Society said:

> Carbon Dioxide Removal methods that remove CO_2 from the atmosphere without perturbing natural systems, and without large-scale land-use change requirements, such as CO_2 capture from air and possibly also enhanced weathering, are likely to have fewer side effects. Techniques that sequester carbon but have land-use implications (such as biochar and soil-based enhanced weathering) may be useful contributors on a small scale although the circumstances under which they are economically viable and socially and ecologically sustainable remain to be determined.

Behind the scenes, as the Royal Society prepared its report, a noisy quarrel that had first erupted in Poznan during the UN's climate negotiations spilled into the deliberations. Soil scientists led by Johannes Lehmann of Cornell University and Peter Read of Massey University in New Zealand told the panel that afforestation provides a short-term response but is limited by land availability, while carbon storage in the soil and BECS could, once scaled up, store the entire excess of atmospheric CO_2 generated by fossil fuels. Read and Lehmann wrote: "Biochar has a key role to play on the bulk of the land areas that will be used for the sustainable production of the biomass raw material, co-produced with food and fiber...."[2]

On the other side of the argument stood a public-interest watchdog group, Biofuelwatch, which warned the panel that biochar was

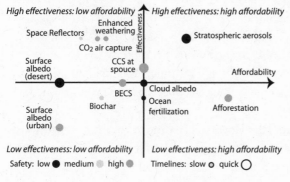

Fig. 33: *The Royal Society concluded that geoengineering could be technically possible, but that there are major uncertainties regarding its effectiveness, costs, governance, and environmental impacts.*

essentially greenwash; overhyped, underresearched, unproven, and dangerous to both the food supply and the atmosphere. The Royal Society adopted Biofuelwatch's position as its own, saying:

[Biochar is] sometimes proposed as an answer to a number of different problems, since it draws down and locks up atmospheric carbon, it can improve crop yields, and it creates biofuels, a renewable energy source. How effectively it achieves each of these goals, at what costs, and with what wider impacts, will determine the influence biochar can have as a geoengineering technology. . . . It remains questionable whether pyrolysing the biomass and burying the char has a greater impact on atmospheric greenhouse gas levels than simply burning the biomass in a power plant and displacing carbon-intensive coal plants. . . .

The residence time of carbon converted to biochar in soils, and the effect on soil productivity of adding large loadings of char is uncertain (Submission: Biofuelwatch). It is known, for example from archaeological sites that charcoal can have a residence time of hundreds or thousands of years in soils. However, the conditions of pyrolysis may affect both the yield of char and its long-term stability in the soil (Submission: U.K. Biochar Research Centre) and further research is required. . . . However, unless the sustainable sequestration rate exceeds around 1 GtC/yr, it is unlikely that it could make a large contribution.

As is the case with biofuels, there is also the significant risk that inappropriately applied incentives to encourage biochar might increase the cost and reduce the availability of food crops, if growing biomass feedstocks becomes more profitable than growing food. Biochar and other forms of sequestered biomass have not yet been adequately researched and characterised, and so should not be eligible for carbon credits . . . until there is a reliable system in place for verifying how much carbon is stored, and the wider social and environmental effects have been determined.

Premises considered, the Royal Society moved biochar over towards the left axis of its chart, signifying reduced safety and greater potential costs than claimed by the advocates.

Thirty

At the Climate Summit in Copenhagen three months later, Biofuelwatch made a push to seal the deal and keep biochar and carbon farming off the short list of options for saving the planet. The group's spokesperson, Deepak Rughani, leafleted delegates, fed questions to public panels, and gave PowerPoint talks on the evils of biochar.[3]

While acknowledging that 14 governments and the UN Convention to Combat Desertification have endorsed biochar, Rughani said that biochar was being fast-tracked without adequate vetting. He said it was like releasing a new pharmaceutical drug without clinical testing. He enumerated several "false claims" of its advocates that he said were going unexamined:

- Carbon-negative cooking and heating. According to Rughani, with any fire, you only get the energy out that you put in, so if you get a third to half out as biochar, that means you have to find a third to half more fuel, which in many parts of the world is already unsustainable and leading to deforestation.
- Elimination of soot. We all know the bad health effects of inhaling soot, endemic in Africa and Asia, and when you add the handling of biochar from stoves, this problem will only worsen.
- Carbon-negative agriculture. We are seeing large clouds of black carbon when biochar is broadcast to the fields. This is only adding to the climate problem, not to say global dimming. When you spread these large swaths of biochar across the ground, you turn the field black. Fine for Japan, where they want to warm the earth, but in Africa the last thing you want is hotter soils.
- Long retention, à la terra preta. Microbial breakdown is what will determine whether the biochar will stay in the soil or not. Up to 72 percent will oxidize and go back to the atmosphere as CO_2 within 20 years. The only place that biochar works to improve crops is in the tropics, and farming the Amazon soils with inorganic carbon would be disastrous for the rainforest.

Rughani argued that, as biochar scavenges soil nutrients and then vaporizes, people will become more dependent on industrial fertilizers, and that will cause more eutrophication of rivers and lakes. He concluded that 156 public-interest organizations have come out against biochar, and that the precautionary principle would suggest that when you undertake geoengineering schemes of this type, more study and vetting of the claims should precede widespread adoption of the practice.

To an uninformed audience, all of that sounds quite reasonable — until you realize that soil carbon has been studied for decades, or in some cases centuries. While advanced studies are still needed and are ongoing, we know enough to conclude that biochar, properly characterized, manufactured, and applied, is ready for prime time. It is a way to get us back below 350 parts per million atmospheric carbon, on decadal rather than millennial timeframes, and is safer by far than the no-action option dictated by the precautionary principle.

While there were some legitimate concerns — all ones that are being addressed by the biochar policy community in a considered and deliberate fashion — there were a huge number of false or misleading statements sprinkled through Rughani's talk. Raising specters of giant tree plantations that displace indigenous societies, an industry that crushes local initiatives, enhanced addictions to fertilizer, and destruction of soil humus on a massive scale, Biofuelwatch pulled no punches.[4]

Although biochar's favorable attributes had already been put forward by soil scientists, producers, permaculturists, and organic farmers at the Climate Summit, the biochar community turned to Nathaniel Mulcahy of WorldStove to address each and every one of Biofuelwatch's arguments. Mulcahy published an open letter refuting the false fears raised by the anti-biochar people, but also acknowledged that legitimate concerns, where they exist, merit attention.

On the issue of biochar tree plantations supplanting agricultural lands, Mulcahy pointed to the landfills of the world overflowing with charable biomass. Egypt produces 20 million tons of rice straw every year; 70 percent of it is burned in the open. In the US, landscape debris comprises more than 32 million tons a year, 13 percent of all solid waste. Globally, aboveground crop residues in 2006 were about 5 billion tons, much of it burned in the field rather than fed to animals or used to improve soils.

Mulcahy wrote that if a third of the original weight of that crop-residue biomass were returned as high-carbon (80-percent) biochar, such as by using efficient stoves or kilns, just 20 percent of current global field wastes could sequester 727,075,555 tons of CO_2 per year without cutting a single tree.

As I write this, Mulcahy is in Haiti working on the 2010 earthquake reconstruction effort by building and deploying stoves — 500 community-scale and 2000 home-size per month. By the end of the year these stoves will be making 20 tons of biochar every day. In preparation for the next phase of the Haiti project, WorldStove has established preliminary agreements with 48 agricultural cooperatives that will provide crop waste for pellet production. The farmers will receive a proportionate amount of biochar in return, which they can use for soil improvement. Mulcahy said that the Haiti experience has confirmed the usefulness of WorldStove's Five-Step program, while adding some new features — such as using biochar for composting toilets, to reduce odors and create a biochar-humanure mixture useful to tree-planters.

Does a stove that produces biochar require more fuel than a stove that does not? WorldStove already has enough field experience to answer that question. Fully pyrolytic stoves produce clean-burning gases from biomass and then burn only those gases. It is much easier to burn gases efficiently than to burn solid fuels. Open fires, used by much of the world, are 7 to 12 percent efficient. Pyrolytic stoves have an efficiency of up to 93 percent, so that they can be used for cooking and heating with less fuel, even while producing biochar for planting drought-hardy forests, and growing vegetables.

Does biochar force farmers to become dependent on fertilizer? Both biochar and humus can contain nutrients when first formed, and those will soon be consumed — but what remains behind has an extraordinary amount of surface area that creates an ideal environment for soil macro- and microflora and fauna. Fertilizer use will diminish with time.

If there is a legitimate concern, it is that biochar and carbon farming will be excluded from the emerging carbon-trading market. In that event, Biofuelwatch's worst nightmares could be realized. In a Wild West scenario — where there are no certification standards, no requirements for life-cycle analysis, no feedstock and product characterization, and no need for continued research on soil biology and plant results — the monocropping, toxicity, displacement of the poor, and all the rest become possible, even probable.

Fig. 34: *Having no machine tools or dyes to use in a disaster recovery zone, Nathaniel Mulcahy modified his stove designs to use available tools, skills, and materials.*

Fortunately, the biochar advocacy community has risen up like the Rocky Mountains to impose order on the commercial process. At the North American Biochar Conference in Boulder, Colorado, in August 2009, roundtables met to discuss criteria for sustainability, how to characterize biochar, and what to require of manufacturers.

Gloria Flora, founder of Sustainable Obtainable Solutions and former US Forest Service forester, introduced draft standards that would compel producers to commit to full life-cycle assessment for energy, water, and carbon footprints. The draft mandates would also require the biochar industry to optimize plant, animal, benthic, and microbial biodiversity, improve forest health and habitat, and assist open and transparent citizen involvement in the construction, operation, and monitoring of facilities and farms. At the conference in Boulder, a policy committee was formed to develop standards for industry and agriculture.[5]

Also at the Boulder conference, David Yarrow, biochar pioneer, small farmer, and permaculture trainer in New York and New England, unveiled a vision of a community-centered biochar lifestyle that obtains fertility, fuel, and food in an ecologically responsible cycle. The three economic drivers for biochar and carbon farming are farm products (including fertilizer, fuels, food, and fiber), climate services, and carbon-negative community. That third driver — the greening of the human habitat to deliver carbon-negative housing and workplaces — means taking a newer, more holistic, look at our built environments.

Carbon trading is coming, whether its opponents want it or not. More regulation of atmospheric pollution is inevitable, and sectors such as electric utilities and transportation will seek ways to offset emissions, with private contracts if not through organized exchanges. Governments will be forced to intervene and regulate, and for biochar, carbon farming, and forest projects, that will be an opportunity, not just to augment financial incentives, but to insist on ethical standards.

Carbon Trading

FOR THE 196 COUNTRIES GATHERED at the UN Conference on Climate Change in Copenhagen in December 2009, carbon trading, cap and share, contraction and convergence, and carbon emissions fees were the subjects of all-night meetings. Climate issues are unquestionably both complicated and profound.

Today the net worth of the world's 358 richest people is equal to the combined income of the poorest 45 percent of the world's population — 2.3 billion people. How did those 358 people obtain their fortunes? At root, they all came from the bowels of the earth, although multigenerational transaction fees concentrated and distilled until they were expressed as ones and zeros in the circuits of electrons moving between banks and national treasuries.

As it accumulated, that accounting maintained its elaborate cyber-ethereal body by sending fossil sunlight back to the heavens as smoke from a billion fires. For the past few centuries, that burning has been saturating the atmosphere with molecules of carbon and other reflective and heat-trapping gases, to the point where the available space for rent is now fully occupied; adding more only oversaturates and alters other balances — sea ice, tropical forests, whale populations, and monsoon rains, among the many.

In Copenhagen, the developing poor asked the industrial rich to pay for that overuse of atmospheric space, and, going forward, to use only an equal allocation — a per-capita ration perhaps — in the future. That entire concept was received with either awkward silence or ridicule. US chief negotiator Todd Stern said:

> I actually completely reject the notion of a debt or reparations or anything of the like. For most of the 200 years since the Industrial Revolution, people were blissfully ignorant of the fact that emissions caused a greenhouse effect. It's a relatively recent phenomenon.[1]

For advocates of biochar, carbon farming, and tree planting, the best case would be to have a firm cap on emissions as part of a binding global treaty, with all nations subscribed. That would put a floor under the international market in carbon trading, which, by the time of the Copenhagen conference, had already become the fastest-growing commodities market on Earth, with more than $300 billion in emission reductions sold.

Anticipating adoption of a floor (say, $25 or $100 per ton of carbon sequestered), Barclays, Citibank, and Goldman Sachs have opened carbon-trading desks in London. Traders from Hong Kong to Brazil — who had speculated on dot-com ventures and commoditized mortgage instruments, and who had only recently propelled crude oil to an economy-wrenching bubble of more than $140 per barrel — had gotten out with fortunes before the crash of all of those sectors, and were queuing up for another, even bigger game. How big? Estimates run to $3 trillion per year. That is a 3 followed by 12 zeros.

The soft-spoken president of the International Biochar Initiative, Cornell University soil scientist Johannes Lehmann, explained to the Copenhagen summit why biochar is not only needed, but possibly is more important than carbon farming or tree planting:

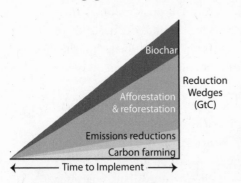

Fig. 35: *While neither the full potential nor implementation time is known for any specific wedge, working together the wedges have to reduce or remove more than 4.5 GtC in excess carbon now accumulating in the atmosphere annually.*

We have the hope that agricultural soil carbon that is non-char will find its way to recognition as an emissions reduction tool, but obviously [bio]char has a number of advantages to make it a bit easier for us to account for the amount of carbon accrual in soil, because we can monitor the amount of [bio]char that is added to soil rather than having to infer the amount of stabilization that happens in soil. The amount of soil carbon from crop residues or leaves from forests is the result of a stabilization mechanism — that organic matter partly decomposes and partly is protected, for instance, within aggregates, on clay surfaces, in small pores, and those stabilization mechanisms are contingent upon the clay content, mineralogy, temperature, moisture, etc., etc. Whereas for [bio]char, those mechanisms are clearly secondary, because [bio]char's stability in soil rests on its chemical recalcitrance. That means that from what you put in, you can predict what will remain. We can relate, from what we know of the chemical recalcitrance, what is put in to what remains in the soil.

Biochar has a huge bonus there — in carbon trading accounting. The second bonus is that we can actually verify that it is there because it is geochemically distinct from other soil elements. We can come back years later and verify that this is the char that we applied on day x.

This could be an advantage that maybe breaks the ice on negotiations on soil carbon sequestration. We can say there are opportunities to do soil carbon sequestration and have it be accountable and not run into issues about permanence and other valid issues. But biochar is not a stand-alone practice. It is part of the agricultural carbon management system. It is part of the whole practice of carbon management, which is important to agriculture, because if you take carbon from one end, then it is missing at the other end, so we have to find a way of accounting for agricultural carbon as a whole, and not take biochar alone separately, apart from the soil carbon equation. It is not one way or another, but looking at the whole.

Of the various mechanisms for regulating emissions being discussed by the UN, the carbon maintenance fee proposed by Ireland's

Foundation for the Economics of Sustainability (Feasta) has perhaps the greatest simplicity and ease of enforcement. The carbon maintenance fee (CMF) would budget world CO_2e on a per-capita basis, charge for emissions, and pay for sequestrations. Surveys indicate that the majority of the populations in industrial countries, even the ones most in denial about climate change, such as the US, Canada, and Australia, favor fines for pollution. It seems most people would agree with John Locke, who said that, when common property is appropriated, there must be "enough, and as good, left in common for others."

The Irish plan would separate the emissions trading, which could be on a modified cap and share basis, from sequestration, which would have its own financial ledger. Landmasses would be surveyed both by satellite imagery and by inspectors taking soil samples and tree counts. If a country lost carbon from poor land management, it would pay a fine. If it gained carbon from better practices, it would receive a payment. The fines would be slightly higher than the payments, which would allow for administration of the program. The expectation is that most countries will take advantage of the incentives to sequester more carbon and earn more rewards, rather than owe taxes.

Feasta's Richard Douthwaite and Corinna Byrne report that research from African, South American, and Asian forests shows that tropical forests remove about 4.8 $GtCO_2e$ from the atmosphere every year, about 18 percent of the annual amount added globally from fossil fuels. African forests alone account for 1.2 Gt. "That could give the countries that have them an annual income of perhaps 6 billion euros plus a carbon maintenance fee," Douthwaite and Byrne say. At $27.20 per ton of carbon, "deforestation could potentially be virtually eliminated."

The elegance of the CMF lies in how it will extend its tendrils down to each state, province, region, municipality, and landholding. To increase the carbon in its soils and earn more money, a country would likely offer incentives to regional governing bodies. Those bodies would turn around and offer incentives to local governments, foresters, ranchers, and farmers. Demand for carbon farming trainers, skilled composters, holistic range managers and foresters, biochar retorts, and keyline plows would blossom and bloom.

The International Biochar Initiative

IN 2007, THE INTERNATIONAL BIOCHAR INITIATIVE BEGAN with a mission to plant the seeds of a sustainable and ethical biochar paradigm, as quickly and cleanly as it could. IBI has also been pushing for biochar's acceptance into the UN Framework Convention on Climate Change (UNFCCC) as one tool for climate change mitigation and adaptation.

Many companies and research laboratories have begun focusing their efforts on biochar pyrolysis or gasification systems, large and small. Others are developing a consumer-friendly biochar product that can sell in retail-size bags in garden stores. Still others are looking at combining these processes with waste reduction to create services they can sell to large woody-waste producers, electric utilities, and carbon traders.

At this writing, the commercialization ramp is still at its lowest elevation, and the companies can only go so far until actual customers show up and want to buy their products. It is a chicken-and-egg situation. You can't sell biochar or its services until you have the manufacturing systems, and you don't need those systems until you have customers to buy the products.

IBI wants to make it possible for new companies to get started making biochar as the customer base gradually expands. It wants both the chicken and the egg, at the same moment. The scientists, engineers, farmers, and students who populate IBI working sessions are not Wall Street investors by and large — they just want to save the planet.

For now, IBI would prefer to stay on safe ground by promulgating some simple ethical guidelines, including:

- First choice in feedstocks should always be a waste stream that if left unattended would either decay or be burned, creating greenhouse gases;
- Quality of air and water counts; and
- Every burn should be a clean burn.

In any biochar production, it is necessary to leave behind biomass residues that feed and replenish the soil, provide seedling shelter, and protect against erosion. Biochar industries need to avoid damaging sensitive lands, sensitive cultures, and sensitive biomes — such as field, stream, and mountain ecologies, and the habitats of endangered plants and animals.

IBI is considering asking its members to commit to full life-cycle assessment, and that includes modeling ERoI, carbon footprint, and resource use, especially of the nonrenewable variety. It has devised effective monitoring plans and urges producers and governments to follow them.

IBI has set up its own free extension service to help researchers, farmers, gardeners, and engineers get the best information. Julie Major is IBI's Agricultural Extension Director, and Jane Lynch is IBI's Technology Extension Director. Together they answer questions, refine standardized tests and classification systems, and gather data from fieldwork in progress on six continents.

For IBI, the cautious approach of the soil scientists, rather than the promotional hyperbole of the industry, has become the dominant style. Johannes Lehmann personally embodies this approach. He is quick to admit what we don't yet know, aware that we are still tampering even when we are trying to restore, and constantly reiterates the need to proceed cautiously, with the best scientific information we can obtain, and try to do no harm. His soft speech, cautious phrasing, understated claims — but abiding sense of urgency, promise, and potential — stand in sharp contrast to the wild rhetorical flourishes of biochar's critics.

What are the relative effectivities of biochar and carbon farming as climate mitigation strategies? A Rodale Institute comparison trial of chemically fertilized versus organic farming showed that soil under organic management in Pennsylvania accumulated about 1000 pounds of carbon per acre-foot of soil each year. This accumulation is equal to about 3500 pounds of carbon dioxide per acre taken from the air and sequestered into soil organic matter. Rodale concluded that converting all US cropland currently planted in corn and soybeans to organic methods would sequester 216 million tons of CO_2e per year, which converts to 0.059 GtC.

The Garnaut Inquiry, looking at carbon farming potential, estimated that Australian soils can sequester, conservatively, 600 million tons of CO_2e per year (0.163 GtC). That is not good enough. Just the US adds nearly 12 times that much — 1.9 GtC — from industry, and those amounts are increasing at 1 percent per year. Rattan Lal's best-case target for carbon sequestration from farming worldwide is 1 GtC/year. So, whether you take Lal's estimate, the Garnaut estimate, or the Rodale estimate, there is no way carbon farming alone could absorb US emissions, to say nothing of dropping atmospheric carbon dioxide concentrations from 390 ppmv to 350 or lower.

We *need* biochar and trees to do that. Even with Lal's estimate of carbon farming potential of 1 GtC/yr globally, being labile the carbon will continue to cycle back to the atmosphere every decade or two. Lehmann and IBI's Jim Amonette estimate that biochar has the near-term potential to sequester 1 GtC/yr (although IBI's Ronald Larson, at the US Biochar Initiative Conference in Iowa in June 2010, suggested that more than 1 GtC/yr might be possible in the US alone). However, biochar has a retention time of 1000 years and so could accumulate fairly quickly. Reforestation, using the step-harvest method described by Frank Michael, in combination with projects like Geoff Lawton's Jordan Valley farm and the Sahara Forest, could sequester 4.5 GtC/yr, withholding that carbon from the atmosphere, potentially, for centuries.

It is important to acknowledge that biochar done in an unsustainable way, or in a way that increases atmospheric carbon loading in the near term, even if reducing it over the long term, is not what we need. We are in the early stages of this agricultural revolution, re-learning how to live in balance. Like a toddler taking first steps, we have not yet found our equilibrium.

34

Permaculture Marines

MARCH 8, 2026, INDIAN OCEAN. The helo lifted Marine Sergeant Vincente Sanchez and his three-member team from the deck of the refitted Vysotsky-class assault carrier *USN David Brower* and banked shoreward over the Gulf of Aden. The landing zone was near Hadaaftimo, Somalia, 10° 45' North, 48° 06' East, in the valley of the Daalo Mountains, where in a massif of solid rock there grows a frankincense tree.

Hadaaftimo is a place where farming ends and desert begins. Up until the early 21st century, this part of the world had been giving up farmland to desert at an unsettling pace, literally. Now the casuistries of the climate gods have altered ocean currents and monsoons and created a shift northward of the mid-African wet zone — fortuitous for Hadaaftimo, but no one can predict how long it might last. The marines are part of an international response network, invited by the new Somali government and working under the United Nations Convention to Combat Desertification (UNCCD). They were trained to exploit just these kinds of opportunities, to enlarge the wedge, to gain as much ground as they can, and then hold it when the climate system moves on.

Lance Corporal Mary Mugané was trained in soil biology at the Western Hemisphere Institute for Security Cooperation at Fort Benning, Georgia. On the ground, she unpacked sample jars and a small scoop and began looking around for the best places to take some soil. She wanted the good stuff, maybe at the base of a damal or qudhac tree, or in a dhirindhir and aristida copse, but she also wanted the worst she could find: the caked mud, the sandy patches where the desert has been establishing, the places the goats had nibbled away the cover.

"Goats," said Lance Corporal Nasid al Homeni, to no one in particular, as he eyed the goats a farmer kept. "I hate goats." He reached for his penetrometer and began taking compaction readings where the goat track entered the field.

"No goats, no feta cheese in the mess," said Marine Corporal Charlie Patterson. Patterson was the team's permaculturist. He was setting up a tripod, breaking out GPS surveying tools, and loosing tiny robotic crawlers and dragonflies that darted across the landscape. Within a few minutes, he would have the topography and orientation of the site on his laptop and be setting up a satellite-linked weather station. Patterson had trained in the N. I. Vavilov Institute of Plant Industry in Saint Petersburg as part of a Russian military exchange and then, with his Russian counterparts, at the American Type Culture Collection in Manassas, Virginia. He was working toward a degree in mycology from Gaia University in his spare time, and this work fit right in with his action-learning track. After setting up the instruments to collect wind and precipitation data, he would scout the ground for mycorrhizal activity.

Sergeant Sanchez had gone to chat with the farmer and his family, who were smiling broadly at the sight of the UNCCD response force and inviting him to come in while they put on tea.

Back aboard the *Brower* that afternoon, the team turned over its samples for lab analysis. The sergeant went off to brief Command about

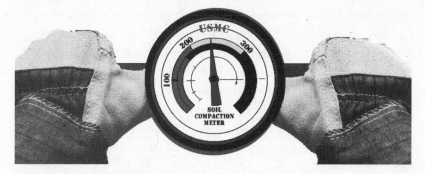

Fig. 36: *Penetrometer soil compaction meter.*

what he had seen and learned on the ground. Using a combination of satellite imagery and special software, Charlie Patterson rendered a holographic map of the Somali landscape and began to plot likely positions for swales, a tree nursery, and windbreaks. He could retrieve some general microclimatic data from NASA, but he would need to leave the weather station in place for a full year or more before he would feel confident he had made the best possible design for that site. He or his successors would be revisiting the location many times to do fine-tuning.

After the lab results came back, Mugané and al Homeni went down to the carrier's brewing bay and gave the Chief Warrant Officer their recipe and the samples of beneficials they had collected. When the tea was ready, the team helo'ed back to the site, along with three cargo-copter loads of "carrier compost" — biochar mixed with their own composted humanure and kitchen wastes, including the earthworm-transformed rinds of Somali goat cheese from the *Brower*'s mess. While they were back aboard, testing and brewing, a team of engineers

Seedballs

In 1938, at a time when industrial agriculture was just taking hold in Japan, Masanobu Fukuoka began saying that the mechanical approach was really not the best idea. Fukuoka, a soil scientist, argued that nature was perfect as it is and that human attempts to try to change the way that nature does something are destined to fail. On his family's citrus farm in Shikoku Island in Southern Japan, Fukuoka reintroduced a traditional technique for broadcasting seed, called *tsuchi dango* (earth dumpling), which consisted of mixing successional seeds of companion plants for the coming seasons with compost, clay, and sometimes manure, and then rolling them with the palm of the hand into small balls and broadcasting the seedballs to the field. The technique is useful for sowing thin and compacted soils, and for avoiding seed eaters. When seasonal rains melt away the clay cover, the seed sets, is fertilized, and grows rapidly. The essence of Fukuoka's method, described in his book, *The One Straw Revolution*, is to reproduce natural conditions as closely as possible: in this case, the action of seabirds, which ingest a seed and deposit it in their rookeries encased in guano. There is no plowing, the seed germinates quite vigorously, and when it is time, the farmer's role is to harvest and reseed.

had been dropped in with light earthmovers, sculpting the ground to the lines of Patterson's virtual grid. The compost copters would follow these lines, laying down mulch on the berms above the wadis. Later, a tactical air strike would bombard the entire area with seedballs.

When Sergeant Sanchez stepped out of the helicopter the next time, he had a special gift for the Somali family. It was a new stove, able to produce a kilo of biochar daily. Now, while they cooked their meals, they could begin earning income from the carbon exchange in Mogadishu and also receive payments from their local CMF authority for contributing to Somalia's net carbon balance.

The *Brower* would remain on station until relieved or reassigned. The world is a big place, and just down the coast is a whole lot more Africa.

The story above is science fiction fantasy, but at its core is a seed of truth. World military spending stands today at roughly $1.5 trillion per year; $742 billion of that is spent by the US, which also has half the world's aircraft carriers, each nearly twice the size of any other nation's largest ship.

Imagine spending that kind of money to actually save the world.

What does defense mean? Does it mean security? Was it a quest for food security that caused Sumer to poison its own soils? Is it a quest for security that drives deserts to the gates of Beijing? What does security really mean now, when humans evolve on million-year timelines, while climate traverses those distances in mere seasons?

If our survival depends on our ability to adapt and evolve, will we? Can we conceive of replacing the military institutions of the world with armies of permaculturists, making biochar and planting trees? And if we can conceive of it, will we do it?

35

Carbon-Negative Communities

THE DESIGN OF COMMUNITIES WILL CEASE, fairly soon, to resemble something plucked from old science fiction magazines or the world's fairs of the 20th century. The bankruptcy of the Dubai World runs deeper than its accounting ledgers. To build cities in the desert with the world's largest shopping malls, highest skyscrapers, and an indoor, air-conditioned ski resort required more than hubris, whimsy, or nearly infinite petrodollars. It required raw ignorance of nature at a previously unimagined level. The future will be different.

What changes for ill in one location may help a change for the better occur somewhere else. The changes needed for humans are adaptive: We need to be able to discern the direction of local and regional trends and to alter our behavior appropriately. From Diamond's list of collapse factors, "societal responses" may turn out to be the most critical. As the British historian Arnold Toynbee said, "Civilizations die from suicide, not by murder."

Approximately half of greenhouse-gas emissions in the industrial world are due to buildings. The low-hanging fruit of retrofits and revised building codes could yield immediate reductions of 60 percent, which translates to 1.35 GtC, equivalent to the savings pledged by governments at the 2010 climate summit in Copenhagen.

In the world to come, the premium will be on buildings and settlement patterns that conserve energy and water; provide food, livelihood, and safety from violent events (natural and man-made); require little

Collapse Factors

Jared Diamond lists a number of factors that have historically contributed to the collapse of past societies (in no particular order):

- Human environmental impacts
  - overpopulation
  - increased per-capita impact of people
  - over-exploitation of natural resources
    - deforestation and habitat destruction
    - soil problems (erosion, salinization, and soil fertility losses)
    - water management problems
    - overhunting
    - overfishing
    - effects of introduced species on native species
- Climate change
- Enemies
- Dependence on trade for some critical imports
- Societal responses
  - economic
  - political
  - social and behavioral

or no nonrenewable resources to build and operate; and are not just carbon-neutral, but carbon-negative. Some of the more daring thinkers in the architectural world are already there.

Long before concepts like "green building" and "sustainability" were fashionable, the husband-and-wife team of Pliny Fisk and Gail Vittori envisioned a future in which architectural design and renewable resources worked together. The nonprofit Center for Maximum Potential Building Systems, which they co-direct, has collaborated on projects as wide-ranging as the eco-friendly renovation of the Pentagon and the development of a model village in Szechuan Province, China.

Sitting under the shade of an oak on the grassy mall outside the Smithsonian Institution, Fisk handed me a cement brick. "It absorbs carbon from the air," he said. It was AshCrete, an innovative building material Fisk had developed from seawater and the waste fly ash generated by coal-fired power plants and aluminum smelters. Most cement-making is a carbon source — about 25 billion tons of carbon have gone to the atmosphere from cement kilns, and production is still growing. Replacing carbon-positive concrete buildings with carbon-negative counterparts is revolutionary.

Peter Harper is head of Research and Innovation at the Centre for Alternative Technology in Wales, UK, where they have created a vision for "Zero-Carbon Britain 2030" that carefully quantifies the stocks and flows of energy, carbon, food, nitrogen, etc. necessary to take the emissions of the British Isles down to zero and below. Natural carbon sequestration processes using soils, forests and biochar are a key contribution, with fully half of the British landscape dedicated to 'carbon crops' used for many purposes. While decarbonizing the UK economy, the scenario delivers higher levels of energy security, food quality and biodiversity, and regenerates the rural economy.

ZeroCarbonBritain predicts that because of the demands of peak oil, financial collapse, and climate change, "by 2027, all Britain's agriculture will be broadly, if not literally, 'organic'." The reasons are fairly obvious. Fifty-four percent of the atmospheric footprint of British agriculture is attributable to nitrogen fertilizer, a potent source of nitrous oxide. The other main components of chemical fertilizer, phosphate and potash, are nonrenewable mineral resources that are quickly depleting. Moreover, the British diet, which favors livestock, is likely to change under a strict carbon-emissions regime. "Despite historical preferences, Britain's citizens are unlikely to spend their entire carbon allowances on beef, mutton, and cheese," the report says.

Harper believes that where space permits, food production will be in and around towns and cities, on allotments, and in private gardens. Sewers will be redesigned to return human wastes, suitably composted, onto the land. The need for more farmland, closer to population centers, will result in what town-planners term "offset density," with areas best suited for agriculture and forestry carved out from suburbia, and residential buildings migrating into tighter clusters. Passive solar technologies such as rooftop water heaters, shading, natural ventilation, greenhouses, and thermal storage walls will join newer concepts such as "bioclimatic" buildings with living roofs, earth berms, and geothermal heat transfer.

Fifteen years ago, Thomas Harttung converted his family farm in Denmark to organic community- supportive agriculture (CSA). Within four years, he was delivering 45,000 boxes of produce to subscriber homes in Denmark, Sweden, and Germany. In 2005, the CSA went carbon-negative with the installation of a CHP burner, pyrolysis kiln, and Stirling engine that supplied 100 percent of needed heat, 50 percent of electricity, and abundant biochar for expanding operations.

In 2007, Jeff Wallin, a green project developer, hired a talented group of environmental consultants to survey some property in Tennessee and draw plans for a sustainable community. His consultants — a new generation of permaculturists, bioregionalists, green engineers, and soil food web cognoscenti led by MaryAnn Simonds — proposed not just a forested subdivision, but a plan for the first carbon-negative eco-community in the world.

The heart of the design consisted of a district heating plant that could be tweaked to maximize production of biochar, biofuels, electricity, or a combination of products, such as activated charcoal, solvents, and chemicals. For the sake of carbon neutrality it would have produced 32,000 tons of biochar per year — 8000 pounds per hour. Around the outskirts of the village were planned continuous-harvest, mixed-species, uneven-aged forestry and fields of switchgrass, elephant grass, and

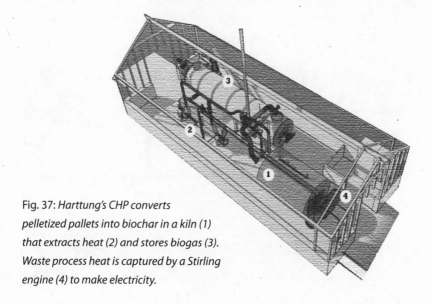

Fig. 37: *Harttung's CHP converts pelletized pallets into biochar in a kiln (1) that extracts heat (2) and stores biogas (3). Waste process heat is captured by a Stirling engine (4) to make electricity.*

bamboo (fertilized with biochar). The project also planned to earn tipping fees from local governments by relieving them of woody biomass that otherwise would have gone to landfills to burn or decompose. Had credits for carbon sequestration been available, Wallin and Simonds' model community would have banked those as well.

Sadly, Wallin and team, EcoTechnologies Group, were forced out of the project by its financier after delivering the business plans and engineering partnerships to make the project a go, and soon after that the Securities and Exchange Commission suspended the project, accusing the financier of defrauding investors.

Wallin was undeterred. He started a new relationship on the island of Kauai in Hawaii, where the group partnered with a business entrepreneur who was reforesting degraded sugar plantation lands. Wallin and partners designed a carbon-negative, zero-waste business model. The principal product will be rare Hawaiian mahogany. Culled "nurse trees" and specialized grasses will be chipped and turned into biochar, heat, and electricity, which will power a cooling facility selling ice, fish, meat, and cold storage. Some of the heat will also go to dry the wood chips and cure finished lumber. Biochar will be sold, used to invigorate the forest, or go to gardens on site. A portion will be fed directly to cattle, horses and fish, to increase the health of the livestock and improve resistance to disease. The project has pre-sold 130,000 verifiable tons of carbon offsets annually from nitrogen fixing mahogany trees, not even counting the biochar.

Fig. 38:
Hawaiian Mahogany farm management plan.

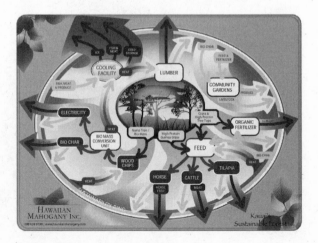

Forty percent of the land area of Japan is called "satoyama," the area between human habitat and wilderness. It is a mosaic of minimal intervention, where farmers, foragers, hunters, and others foray, take a little out, and leave it alone to regrow. Satoyama is a mixture of grasslands, hardwood, softwood, bamboo, and semi-wild copse. In recent years, it has become a nuisance, because as farmers have abandoned the countryside, the bamboo groves have overgrown and wild boars have begun rampaging rice paddies. Now a new idea, "cool farming" cooperatives, has emerged to reinvigorate the rural economy and restore the satoyama ecosystems. Bamboo is being harvested for biochar, the biochar returned to the farms for soil health and carbon credits, and the produce is sold as carbon-negative "cool food." In 2009, the first "cool" cabbage was processed into slaw by a supermarket chain, sold at a premium price and it sold out. Cool Slaw and other carbon negative products may represent a new way to revive rural economies while redeeming ecological services.[1]

Well before the crisis of 2008 struck the economy of Ireland, a group of young Dubliners began looking to the countryside for a place to shelter from the peak oil storm. Criteria: good farmland, a safe water supply, a railroad connection, and a friendly culture. After narrowing the candidates, they chose Cloughjordan in County Tipperary.

Cloughjordan Ecovillage is both a novel experiment and a prod to the older village of Cloughjordan to evolve itself into a cooperative, self-reliant community from the bottom up. The 67-acre development now includes 132 households, community gardens, and an extensive area being returned to woodland. A CSA provides groceries produced to demand, and neighboring farms are sowing fields of grasses that will supply pellets for the central heating plant that will not only warm the ecovillage in winter, but replace coal and peat burning in much of the older town.

Design and planning involves all ecovillage members and is guided by permaculture principles. Going carbon-negative, while not imagined in the first phase, is now within easy reach, requiring only the addition of a pyrolyzing kiln to the pellet heater, and management of the woodlands through step-harvest and Pioneer Forest techniques.

These examples of carbon-negative communities are from the industrial world, where the distance to go is farthest and the carbon debt greatest, but examples of similar lifestyle and built-environment change can also be found in the global South, from Auroville in India to Gaviotas in Colombia. Welcome to the future. If our grandchildren survive the 21st century, this will be their world.

In 1902, British master planner Ebenezer Howard published a small tome entitled *Garden Cities of To-Morrow*, which argued that by the end of the 20th century the factory metroplexes of England, smelling of burnt coal and horse manure, would be replaced by medium-sized "garden cities" designed to fit into their landscapes and supply their own food while nourishing the souls of their inhabitants with clean air, tree-lined sidewalks, and convivial apportionment of urban space. It was fully a century later that archaeologist Michael Heckenberger discovered Howard's garden cities not in London but in the central Amazon, where Orellana's expedition had reported them. Depending on how you reckon it, Howard was off by either 50 or 500 years.

Assume that, all of a sudden, we were to awaken to the threat posed by conventional agriculture to our survival. What changes in our arrangements might, even at this late hour, offer some hope?

It would likely involve some combination of biochar, carbon farming, tree planting, and redesign of the built environment and energy systems to be carbon-negative. I cannot imagine any alternative that excludes those strategies that would remain viable for very long.

Transition is its own challenge. Existential threats are not unprecedented in the history of our genome, and that provides some comfort. We made it through all the evolutionary bottlenecks we know of, or we wouldn't be here now. Over the course of our evolution we have benefited from stable climate and dense biodiversity. That biodiversity has given us, with our linear thought limitations, a safe refuge within the nonlinear web of life that indefatigably minds the store when we are out to lunch.

This stability is something we will soon have a lot less of, and adjusting to the suddenness of changed circumstances will likely become our greatest challenge.[2]

Roughly one-third of the land surface of the planet is occupied by human settlement in various stages of development. Another third is forest, which is shrinking, and the remainder is desert, which is expanding. All of the world's deserts at one time supported vegetation and would do so again if the freshwater and respite were available. Water has become critically important for our future, and to combat climate change we will need to expand forests, which preserve freshwater, and to hold back the deserts.

Ironically, the near-extinction of Native Americans provided a sliver of evidence that could prod us from a completely unsustainable habitation pattern to one that offers not only the prospect of continuation of the human experiment but the possibility of effectuating repair to Gaia's tattered systems. The signal, glimpsed from ice cored in Greenland, tells us that it is more probable than not that the Medieval Maximum and the Little Ice Age came at least in part from the rise and fall of the great pre-Columbian city-states. That new knowledge can inspire us as we begin planting trees and making terra preta soils.

While the Sumerians, Babylonians, Assyrians, and Macedonians were busy salting the Fertile Crescent, the indigenous peoples of the Americas were quietly building societies in which conservation was a consequence of production. Ronald Nigh discovered one when he went into the Petén to study the Lakandon milpas. Like other methods of sustainable agriculture, the Lakandon's is based on observing productive natural ecosystems and mimicking the processes and relationships that make them more resilient and regenerative.

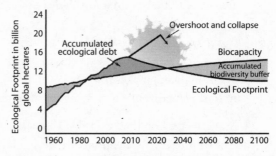

Fig. 39: *If we were to continue with business as usual, by mid-century we would need more than 30 Earths to meet our resource needs, which is impossible. By 2010, we had reached a decisional tipping point.*

Now, with biochar and carbon farming, we can add biocapacity to provide time to move out of overshoot. However, unless we change the industrial growth trend, collapse cannot be averted. After World Wildlife Federation et al., Living Planet Report 2006.

Geoff Lawton is fond of saying that all of the world's problems can be solved in the garden. But the challenge is not to garden; it's not even the need to switch broad-scale agriculture from "conventional" to organic. The greatest challenge is in changing the growth paradigm away from Civilization 1.0 — fossil-fueled industrial globalization — to Civilization 2.0 —"glocalization," a steady-state economy characterized by stable (and in the near term declining) population size.

We don't know exactly what our future built environments will look like, but we can make some educated guesses. They are unlikely to look like today's steel-grey megacities and asphalt highways. More likely, they will resemble the description Father Gaspar de Carvajal, scribe to Francisco de Orellana, made in 1542:

> That was a wonderful thing to see: Even though we went past fleeing and could not examine the place or know what was there, or anything of the peoples in those regions, according to the disposition of what we saw, the interior must be populated much as what we had seen, and thus, this area . . . we must say is "grandísimo."[3]
>
> There were many roads here that entered into the interior of the land, very fine highways . . . there could be seen some very large cities that glistened in white. This, the land, is as good, as fertile, and as normal in appearance as our Spain . . . where much wheat may be harvested and all kinds of fruit trees may be grown. Besides this, it is suitable for the breeding of all sorts of livestock, because on it there are many kinds of grass, just as in our Spain, such as wild marjoram and thistles of a colored sort and scored, and many other very good herbs. The woods of this country are groves of evergreen oaks and plantations of cork trees bearing acorns (for we ourselves saw them) and groves of hard oak. The land is high and makes rolling savannas, the grass not higher than up to the knees, and there is a great deal of game of all sorts.[4]

That passage could as well describe the vista that now stretches in front of us; the path we can take if we so choose: carbon-negative cities, practicing sustainable, bioregionalist agriculture, in a landscape of cultivated ecologies, in millennial balance with the orbit and precession of Earth.

Notes

Chapter 1: The Roots of a Predicament
1 Spiro Kostof, *A History of Architecture: Settings and Rituals* (New York Oxford: Oxford Univ. Press USA, 1985).

Chapter 2: Sombroek's Vision
1 Charles C. Mann, "A Few Words About Wim Sombroek," in William I. Woods et al., eds., *Amazonian Dark Earths: Wim Sombroek's Vision* (New York: Springer, 2009).

2 Wim Sombroek, "Biomass and Carbon Storage in the Amazonian Ecosystems," *Interciencia* 17 (1992): 269–72.

Chapter 3: Conquistadors
1 Gavin Menzies, *1421: The Year China Discovered America* (New York: Harper Collins, 2006); Gavin Menzies, *1434: The Year a Magnificent Chinese Fleet Sailed to Italy and Ignited the Renaissance* (London: Harper Collins, 2008).

Chapter 4: El Dorado
1 Menzies, *1421: The Year China Discovered America.*

2 *Gaspar de Carvajal, Descubrimiento del Río de las Amazonas: Relación de Fr. Gaspar de Carvajal (1584), Exfoliada de la Obra de José Toribio Medina, Edición de Sevilla, 1894 por Juan B. Bueno Medina,* (Madrid: Biblioteca Virtual Miguel de Cervantes), cervantesvirtual.com/servlet/SirveObras/23586289872392741254679/ p0000001.htm#I_0_ (accessed June 13, 2010), translated by the author Albert Bates for *The Biochar Solution.*

3 Assuming Orellana used the same league distance used by Columbus, a league is 3.07 miles.

4 Carvajal, *Descubrimiento del Río de las Amazonas.*

5 José Toribio Medina, *The Discovery of the Amazon, According to the Account of Friar de Caravajal and Other Documents* (1894), English translation by Bertram T. Lee, ed. by H. C. Heaton, (New York: Amer. Geographical Soc., 1934), reprinted in

Vilhjalmur Stefansson, *Great Adventures and Explorations: From the Earliest Times to the Present as Told by the Explorers Themselves* (New York: Dial Press, 1947).

Chapter 5: The Great White Way

1 Carvajal, *Descubrimiento del Río de las Amazonas.*
2 Medina, *The Discovery of the Amazon.*
3 Medina, *The Discovery of the Amazon.*
4 Carvajal, *Descubrimiento del Río de las Amazonas;* Medina, *The Discovery of the Amazon.*
5 See, e.g., Betty J. Meggers, "Environmental Limitation in the Development of Culture," *American Anthropology* 56 (1954): 801–23; and Betty J. Meggers, *Amazonia: Man and Culture in a Counterfeit Paradise* (Chicago: Aldene-Atherson, 1971); but see also Charles C. Mann, "1491," *The Atlantic* (March 2002), theatlantic.com/doc/200203/mann (accessed June 13, 2010).
6 One 17th Century German Jesuit missionary undertook to describe the flora and fauna in the region around Santarem and in so doing provided the earliest known use of the term "terra preta." John Phillip Bettendorff, *Chronicle of the mission of priests of the Society of Jesus in the State of Maranhão* (1690). Belém: Fundação Cultural Tancredo Neves, SECULT, 1990.

Chapter 6: The View from the Bluff

1 Menzies, *1421: The Year China Discovered America.*
2 Lilian Rebellato, William I. Woods, and E. G. Neves, "Pre-Columbian Settlement Dynamics in the Central Amazon," in William I. Woods et al., eds., *Amazonian Dark Earths: Wim Sombroek's Vision* (New York: Springer, 2009).
3 Charles R. Clement, "1492 and the Loss of Amazonian Crop Genetic Resources. I. The Relation between Domestication and Human Population Decline," *Economic Botany* 53 (1999): 188–202.
4 E. G. Neves et al., "The Timing of Terra Preta Formation in the Central Amazon: New Data from Three Sites in the Central Amazon," in Johannes Lehmann et al., eds., *Amazonian Dark Earth: Origin, Properties, Management* (Dordrecht: Kluwer Academic Publishers, 2001).
5 Santiago Mora, Luisa Fernanda Herrera, and Ines Davelier, "Cultivars, Anthropic Soils, and Stability: A Preliminary Report of Archaeological Research in Araracuara, Colombian Amazonia," Univ. of Pittsburgh, *Latin American Archaeology Reports* 2 (1991).
6 Charles C. Mann, *1491: New Revelations of the Americas Before Columbus* (New York: Knopf, 2005).
7 William M. Denevan, ed., *The Native Population of the Americas in 1492* (Madison, Wisconsin: University of Wisconsin Press, 1992).
8 Betty J. Meggers, "Prehistoric Population Density in the Amazon Basin," in John W. Verano and Douglas H. Ubelaker, eds., *Disease and Demography in the Americas* (Washington, DC: Smithsonian Institution Press, 1992), 197–206.
9 T. P. Myers, "El Efecto de las Pestes sobre las Poblaciones de la Amazonía Alta," *Amazonía Peruana* 8 (1988): 61–81.
10 Clement, "1492 and the Loss of Amazonian Crop Genetic Resources."

11 University of New Hampshire tropical ecologist Michael Palace has been awarded a $364,000 grant from NASA's Space Archeology program to estimate the population of pre-Columbian indigenous peoples in the Amazon Basin lowlands by means of satellite remote sensing technology. .

12 Ugo A. Perego, Norman Angerhofer, Maria Pala, et al., The Initial Peopling of the Americas: A Growing Number of Founding Mitochondrial Genomes from Beringia, *Genome Res.* published online June 29, 2010, doi: 10.1101/gr.109231.110 (accessed July 2, 2010).

13 Charles R. Clement, Joseph M. McCann, and Nigel J. H. Smith, "Agrobiodiversity in Amazônia and Its Relationship with Dark Earths," in Johannes Lehmann et al., eds., *Amazonian Dark Earth: Origin, Properties, Management* (Dordrecht: Kluwer Academic Publishers, 2001), 159–78.

14 Richard Nevle, visiting scholar in Dept. of Geological and Environmental Sciences at Stanford University, and Dennis Bird, professor in Dept. of Geological and Environmental Sciences at Stanford University, presented this hypothesis at the annual meeting of the American Geophysical Union on December 17, 2008.

Chapter 7: Confederados

1 Justin Horton, "The Second Lost Cause: Post-National Confederate Imperialism in the Americas" (master's thesis, East Tennessee State Univ. Dept. of History, 2007).

2 Ballard S. Dunn, *Brazil, the Home for Southerners: or a Practical Account of what the Author, and Others, who Visited that Country, for the Same Objects, Saw and Did While in that Empire* (New York: J. B. Richardson, 1866).

3 William I. Woods and William M. Denevan, "Amazonian Dark Earths Research: Initial Reports," US Biochar Initiative Conference, Ames, IA, 2010; biorenew.iastate.edu/events/biochar2010/conference-agenda/agenda-overview. html (accessed July 2, 2010).

4 William I. Woods and William M. Denevan, "Amazonian Dark Earths: The First Century of Reports," in William I. Woods et al., eds., *Amazonian Dark Earths: Wim Sombroek's Vision* (Berlin: Springer, 2009).

Chapter 8: Hartt's Breakthrough

1 William R. Brice and Silvio F. de M. Figueirôa, "Charles Frederick Hartt — A Pioneer of Brazilian Geology," *GSA Today* (March 2003).

2 Woods and Denevan, "Amazonian Dark Earths."

3 Emma Marris, "Putting the Carbon Back: Black is the New Green," *Nature* 442 (10 August 2006): 624–26.

4 Woods and Denevan, "Amazonian Dark Earths."

5 Joshua Hammer, "The Price of Paradise," *The New York Review of Books* (8 October 2009).

Chapter 9: City Z

1 David Grann, *The Lost City of Z: A Tale of Deadly Obsession in the Amazon* (New York: Vintage Departures, 2010).

2 Grann, *The Lost City of Z.*

Chapter 10: Making Sand
1 Jacques Cauvin, *The Birth of the Gods and the Origins of Agriculture* (Cambridge: Cambridge Univ. Press, 2000), originally published in French as *Naissance des Divinités, Naissance de l'Agriculture* (Paris: C.N.R.S. Éditions, 1994).
2 Cauvin, *Birth of the Gods*, 51–61; and Jared Diamond, *Guns, Germs, and Steel: The Fates of Human Societies* (New York: W. W. Norton & Company, 1997).
3 Steven Mithen, *After the Ice: A Global Human History 20,000–5000 BC* (London: Weidenfeld & Nicolson Ltd., 2003), 4.
4 Jim Marrs, *Rule By Secrecy: The Hidden History that Connects the Trilateral Commission, the Freemasons, and the Great Pyramids* (New York: HarperCollins, 2000).
5 Jared Diamond, *Collapse: How Societies Choose to Fail or Succeed* (New York: Viking Press, 2005).

Chapter 12: Changing the Paradigm
1 About 12.8 billion acres (5.2 billion hectares) according to the United Nations Convention to Combat Desertification (UNCDD).
2 N.J. Middleton and D.S.G. Thomas (eds.), *World Atlas of Desertification* New York: United Nations Environmental Programme (1997).

Chapter 13: The Amazon and the Ice Age
1 W. F. Ruddiman, *Plows, Plagues and Petroleum: How Humans Took Control of Climate* (Princeton and Oxford: Princeton Univ. Press, 2005).
2 David Roffe, *Domesday: The Inquest and The Book* (Oxford: Oxford University Press, 2000), 224–49.

Chapter 14: Predicting Climate's Meander
1 Allan J. Yeomans, *Priority One: Together We Can Beat Global Warming* (Arundel, Queensland: Biosphere Media LLC, 2007).
2 US Advisory Committee on Global Change Research, *Global Climate Change Impacts in the United States* (Cambridge: Cambridge Univ. Press, 2009).

Chapter 15: Carbon Farming
1 Rattan Lal and Ronald F. Follett, eds., *Soil Carbon Sequestration and the Greenhouse Effect* (Madison: Soil Science Society of America, 2009).
2 Others put the soil potential as much higher, and moreover, there could be an ocean uptake potential of 50 GtC/yr; See: Ronal Larson, "Biochar GHG Reduction Accounting in Potential Biochar Greenhouse Gas Reductions," US Biochar Initiative Conference, Ames, IA, 2010.
3 James Bruges, *The Biochar Debate: Charcoal's Potential to Reverse Climate Change and Build Soil Fertility* (US: Chelsea Green, 2010), 79.
4 Johannes Lehmann, John Gaunt, and Marco Rondon, "Bio-Char Sequestration in Terrestrial Ecosystems — A Review," *Mitigation and Adaptation Strategies for Global Change* 11, no. 2 (March 2006): 403–27.

Chapter 16: Understanding Soil
1 F. H. King, *Farmers of Forty Centuries: or Permanent Agriculture in China, Korea and Japan* (Emmaus: Rodale Press, 1990), 54.

2 Production of the most commonly used biofuels in the European Union (ethanol made from wheat and maize, and diesel made from rapeseed) and in North America (ethanol made from maize) could increase greenhouse gas emissions, depending on land use and crop management, scale, and other factors. Edward M. W. Smeets et al., "Contribution of N_2O to the Greenhouse Gas Balance of First-Generation Biofuels," *Global Change Biology* 15, no. 1 (2009): 1–23.

3 Union of Concerned Scientists, *Industrial Agriculture: Features and Policy* (Cambridge, Mass., 2007), ucsusa.org/food_and_agriculture/science_and_impacts/impacts_industrial_agriculture/industrial-agriculture-features.html (accessed June 13, 2010).

4 David Pimentel et al., "Impact of Population Growth on Food Supplies and Environment," *Population and Environment* 19, no. 1 (September 1997): 9–14.

5 A. Bouwman et al., *Global Estimates of Gaseous Emissions of NH_3, NO and N_2O from Agricultural Land* (International Fertilizer Association, Food and Agriculture Organization of the United Nations, 2001); Charles Rice, "Introduction to Special Section on Greenhouse Gases and Carbon Sequestration in Agriculture and Forestry," *Journal of Environmental Quality* 35 (2006): 1338–40.

6 United States Environmental Protection Agency, *Global Mitigation of Non-CO_2 Greenhouse Gases (EPA Report 430-R-06-005)* (Washington, DC: EPA, 2006), epa.gov/nonco2/econ-inv/international.html (accessed June 13, 2010).

7 United States Environmental Protection Agency, *Global Anthropogenic Non-CO_2 Greenhouse Gas Emissions: 1990–2020 (EPA Report 430-R-06-003)* (Washington, DC: EPA, 2006).

8 Secretariat of the United Nations Convention to Combat Desertification, *Sustainable Land Management Techniques Related to Climate Change Mitigation/Adaptation and Desertification* (North American Biochar Conference, 2009).

9 Use of the term "ERoI" traces to the Ph.D. thesis on fish migration of Charles A. Hall, a student of Howard T. Odum. Both Joseph Tainter and Thomas Homer-Dixon concluded that a falling ERoI in the Later Roman Empire was one of the reasons for the collapse of the Western Empire in the fifth century CE.

Chapter 19: Compost

1 William Bryant Logan, *Dirt: the Ecstatic Skin of the Earth* (New York: W. W. Norton, 2006), 43–44.

2 Glimus intradices, G. mosseae, G. aggregatum, G. etunicatum (5900 prop./lb each); G. clarum, G. deserticola, Gigaspora margarita, G. Brasilianum, G. Monosporum (1135 prop./lb each).

3 Rhispogen villosullus, R. luteolus, R. amylopohgon, R. fulviglega (95 million prop./lb each); Pisolithus tinctorius (568 million prop./lb); Laccaria bicolor and L. laccata (38 million prop./lb each), Scleroderma cepa, S. citrinium (198 million prop./lb each); Sullius granulatas and S. punctatapies (118 million prop./lb. each).

4 Bacillus licheniformis, B. azotoformans, B. megaterium, B. coagulans, B. pumlis, B. thuringiensis, B. stearothermiphilis, Paenibacillus poxymyxa, P. durum, P. florescence, P. gordonae, Azotobacter polymyxa, A. chroococcum, Sacchromyces cervisiae, Pseudomonas aureofaceans (372 million cfu/lb each).

5 Trichoderma harzianum and T. konigii (150 million cfu/lb each).

6   Lisa M. Hamilton, *Shumei Natural Agriculture: Farming to Create Heaven on Earth, Part III, The Evolution of Natural Agriculture: Not a Farming Method, but a Shared Vision for a Better World* (Emmaus, PA: Rodale Institute, 2003), newfarm.rodaleinstitute.org/international/features/0903/shumei3/shumei3.shtml (accessed June, 13 2010).

7   Rudolf Steiner, *Spiritual Foundations for the Renewal Of Agriculture: A Course of Lectures Held at Koberwitz, Silesia, June 7 to June 16, 1924)* (Kimberton, Penn.: Bio-Dynamic Farming and Gardening Association, 1993).

Chapter 20: Tea Craft and Designer Biochar
1   Elaine Ingham, *Compost Tea Quality: Light Microscope Methods* (Corvallis, Oregon: Soil Foodweb Inc., 2004).

Chapter 21: From Biochar to Terra Preta
1   Bruno Glaser, "Prehistorically Modified Soils of Central Amazonia: A Model for Sustainable Agriculture in the Twenty-First Century," *Philosophic Transactions of the Royal Society of London Series B* 362 (27 February 2007): 187–196.

2   Johannes Lehmann, *Terra Preta de Indio* (Cornell University Dept. of Crop and Soil Sciences), citing Sombroek 1966, Smith 1980, Kern and Kämpf 1989, Sombroek et al. 1993, Woods and McCann 1999, and Glaser et al. 2000, css.cornell.edu/faculty/lehmann/research/terra%20preta/terrapretamain.html (accessed June 13, 2010).

3   Julie Major, *Guide to Conducting Biochar Trials* (International Biochar Initiative, 2009), biochar-international.org/extension; Hugh McLaughlin, *Characterizing Biochars Prior to Addition to Soils — Version I* (Alterna Biocarbon Inc., Jan 2010).

4   David W. Rutherford, Colleen E. Rostad, and Robert L. Wershaw, *Effects of Formation Conditions on the pH of Switchgrass Biochars* (Denver: US Geological Survey, 2009), cees.colorado.edu/docs/characterization/Rutherford%20Poster.pdf (accessed June 13, 2010).

5   Jean-François Ponge et al., "Ingestion of Charcoal by the Amazonian Earthworm *Pontoscolex corethrurus:* A Potential for Tropical Soil Fertility," *Soil Biology and Biochemistry* 38 no. 2006 (2009): 2008–09.

Chapter 22: Making Charcoal
1   André Seidel, *Charcoal in Africa: Importance, Problems and Possible Solution Strategies* (Eschborn: Deutsche Gesellschaft für Technische Zusammenarbeit (GTZ) GmbH, Household Energy Programme — HERA, 2008), gtz.de/de/dokumente/gtz2008-en-charcoal-in-africa.pdf (accessed June 13, 2010).

2   Burkhard Bilger, "Hearth Surgery," *The New Yorker* (December 21, 2009): 84–97.

Chapter 23: Stove Wars
1   Information re the Adam Retort, biocoal.org/3.html (accessed June 13, 2010).

2   US Dept. of Agriculture Agricultural Research Service, "Pyrogasification of Dewatered Swine Solids to Produce Combustible Gas and Biochar," research project, ars.usda.gov/research/projects/projects.htm?ACCN_NO=418265 (accessed June 13, 2010).

3   0.1095 tons times 80% is 0.0876 tons per year of carbon; times 2.86 (C=12, O=16) yields 0.25 tons of CO_2.

Chapter 24: Milpas

1 James D. Nations and Ronald B. Nigh, "The Evolutionary Potential of Lacandon Maya Sustained-Yield Tropical Forest Agriculture," *Journal of Anthropological Research* 36 no. 1 (1980): 1–30; Ronald B. Nigh, "Trees, Fire and Farmers: Making Woods and Soil in the Maya Forest," *Journal of Ethnobiology* 28 no. 2 (2008) 231–243.

2 Robert Jensen, "Sustainability and Politics: An Interview with Wes Jackson," *Counterpunch* (July 10, 2003); Cindy M. Cox et al., "Prospects for Developing Perennial Grain Crops," *Bioscience* (August 2006): 649–59.

3 William M. Denevan, *Cultivated Landscapes of Native Amazonia and the Andes* (Oxford: Oxford Univ. Press, 2001).

4 Albert Bates, "Going Deep in Belize," *The Permaculture Activist* 71 (Spring 2009).

5 World Agroforestry Centre, "Trees on Farms Key to Climate and Food-Secure Future," press release (Nairobi: July 24, 2009); Kate Trumper et al., *Natural Fix? The Role of Ecosystems in Climate Mitigation: A UNEP Rapid Response Assessment* (Cambridge: UNEP-WCMC, 2009).

Chapter 25: Chinampas

1 Phil Crossley, "Just Beyond the Eye: Floating Gardens in Aztec Mexico," *Historical Geography* 32 (2004): 111–135; Phil Crossley, "Sub-Irrigation in Wetland Agriculture," *Agriculture and Human Values* 21 (2004): 191–205.

2 Gary Clyde Hufbauer and Jeffrey J. Schott, *NAFTA Revisited: Achievements and Challenges* (Washington, DC: Institute for International Economics, October 2005) 283–363.

Chapter 26: Trees

1 Jed W. Fahey, "Moringa oleifera: A Review of the Medical Evidence for Its Nutritional, Therapeutic, and Prophylactic Properties," *Trees for Life Journal* 1 (2005): 5.

Chapter 27: The Power of Youth

1 Frank Michael, email communication with the author, March 1, 2010.

Chapter 28: Greening the Desert

1 The Permaculture Research Institute of Australia, "Jordan Valley Permaculture Project," permaculture.org.au/project_profiles/middle_east/jordan_valley_permaculture_project.htm (accessed June 13, 2010).

Chapter 29: Sahara Forest

1 The Bellona Foundation, "The Sahara Forest Project," saharaforestproject.com (accessed June 19, 2010).

Book V: At the Turning Point

1 Timothy M. Lenton et al., "Tipping Elements in the Earth's Climate System," *Proceedings of the National Academy of Sciences* 105 no. 6 (2008): 1786–1793, pnas.org/cgi/doi/10.1073/pnas.0705414105 (accessed June 13, 2010); H. J. Schellnhuber, *Terra Quasi-Incognita: Beyond the 2° C Line* (Oxford: Pottsdam Institute for Climate Impact Research, 2009), PowerPoint presentation, 4 Degrees and Beyond International Climate Conference.

Chapter 31: The Biochar Critique

1   The Royal Society, *Geoengineering the Climate: Science, Governance and Uncertainty* (London: The Royal Society, 2009), royalsociety.org/Geoengineering-the-climate/ (accessed June 13, 2010).

2   Johannes Lehmann and Stephen Joseph, eds., *Biochar for Environmental Management: Science and Technology* (New York: Earthscan, 2009).

3   Almuth Ernsting and Rachel Smolker, *Biochar for Climate Change Mitigation: Fact or Fiction?* (February 2009), biofuelwatch.org.uk/docs/biocharbriefing. pdf (accessed June 13, 2010); and see: Gary DeLong, "Agriculture Soil Offset Standard," US Biochar Initiative Conference, Ames IA, 2010.

4   Deepak Rughani's 16-minute talk at United Nations Climate Change Conference, Cophenhagen, 2009, is available as an audio file atthegreatchange. com/gcdownloads/DeepakAntiBiochar.mp3 (accessed June 13, 2010).

5   One early product of this work is Keith Driver and John Gaunt, *Bringing Biochar Projects into the Carbon Marketplace: An Introduction to Carbon Policy and Markets, Project Design, and Implications for Biochar Protocols* (Carbon War Room et al., 2010), biocharprotocol.org (accessed June 13, 2010).

Chapter 32: Carbon Trading

1   Andrew C. Revkin and Tom Zeller Jr., "US Negotiator Dismisses Reparations for Climate," *The New York Times* (December 10, 2009).

Chapter 35: Carbon-Negative Communities

1   Steven McGreevy, "In a Critical Countryside: Biochar's Foray into Eco-Branding and Japanese Rural Revitalization," US Biochar Initiative Conference, Ames, IA, 2010.

2   Johan Rockström et al., "Planetary Boundaries: Exploring the Safe Operating Space for Humanity," *Ecology and Society* 14 no. 2 (2009): 32.

3   Carvajal, *Descubrimiento del Río de las Amazonas.*

4   Medina, *The Discovery of the Amazon.*

Index

About the Author

A FORMER ENVIRONMENTAL RIGHTS LAWYER, paramedic, brick mason, flour miller, and horse trainer, Albert Bates shared the Right Livelihood Award in 1980 as part of the steering committee of Plenty, working to preserve the cultures of indigenous peoples, and board of directors of The Farm, a pioneering intentional community in Tennessee for the past 39 years. A cofounder and past president of the Global Ecovillage Network, he is presently GEN's representative to the United Nations climate talks. When not tinkering with fuel wringers for algae, hemp cheeses, or pyrolizing cookstoves, he teaches permaculture, ecovillage design and natural building. He tweets at @peaksurfer and blogs at peaksurfer. blogspot.com.